Novel Photochemical Reactions in Organic and Medicinal Chemistry

This book deals with chemical reactions which are triggered when light energy is absorbed by a substance's molecules. It highlights the synthesis, molecular mechanism, chemical and biological properties of organic compounds and also focuses on recent advances in drug development related to cancer and other ailments. It will be helpful for researchers during target identification, lead identification, and optimization during the drug development process as well as in designing new molecules based on the data retrieved from the literature survey.

Features:

- Methods for creating carbon-carbon and carbon-heteroatom bonds following green chemistry and concepts on electron-donor-acceptor pathways.
- Focuses on medicinally important molecules, in particular against diverse cancers.
- Describes synthesis of natural products.
- Describe economical and practical pathways.
- Encourages new ways to conduct research in organic and medicinal chemistry.

This book will be beneficial to undergraduates, postgraduates, research scholars, and young teachers in chemistry and biological science, pharmacists, pharmacologists, clinicians, industrialists, and biochemists.

Novel Photochemical Reactions in Organic and Medicinal Chemistry

By

Aparna Das and Bimal Krishna Banik

CRC Press

Taylor & Francis Group

Boca Raton London New York

CRC Press is an imprint of the
Taylor & Francis Group, an **informa** business

First edition published 2026
by CRC Press
2385 NW Executive Center Drive, Suite 320, Boca Raton FL 33431

and by CRC Press
4 Park Square, Milton Park, Abingdon, Oxon, OX14 4RN

CRC Press is an imprint of Taylor & Francis Group, LLC

© 2026 Taylor & Francis Group, LLC

ISBN: 9781041061946 (hbk)
ISBN: 9781041061953 (pbk)
ISBN: 9781003634249 (ebk)

DOI 10.1201/9781003634249

Typeset in Times
by KnowledgeWorks Global Ltd.

Contents

Contents

About the authors

Dr. Aparna Das

Aparna Das earned her Ph.D. degree in Material science/nanophysics from Joseph Fourier University based on her work on "Semiconductor Quantum Dots for Opto-chemical Sensor Application" at French Alternative Energies and Atomic Energy Commission (CEA), France. Her postdoctoral experiences include working as a research scientist at the California Nano Systems Institute and Electric Engineering Department, University of California, Los Angeles (UCLA), USA, and as an experienced researcher at the Nanowiring-Marie Curie Initial Training Network, Georg August Universität Göttingen, Germany. She also worked as an Assistant Professor at Prince Mohammad Bin Fahd University, Saudi Arabia. Currently, she is a Principal Scientist at Kertz Ltd, UK.

Dr. Das has published more than 145 peer-reviewed articles including 4 books (Elsevier, CRC Press, PMU press), 58 book chapters, 5 press release articles, 4 editorials, and 71 journal publications. She has 53 contributions to international conferences. She has received several conference invitations to present her research work. Importantly, a significant portion of her research contributions are undergoing editorial reviews by many journals. Dr. Das has served as a guest editor of an international journal and she has acted as a reviewer. Dr. Das has obtained several competitive grants and has filed a patent application. She has received the young women researcher award, Outstanding Researcher Award, European microscopy society (EMS) outstanding paper awards and several international fellowships such as Marie-Curie fellowship, CEA-CNRS research fellowship and Brain Korea 21 fellowship. Dr. Das' research works have been recognized by Bentham publisher (Press release at the AAAS Site), Heidelberg University, and Prince Mohammad Bin Fahd University.

Dr. Das' research interests include Computer-assisted Physico-chemical methods, quantum mechanical calculations, interdisciplinary science (Biology, Chemistry, and Physics) for drug development, computer-aided drug design, microwave applications, photochemical reactions, III-Nitride-based chemical sensors, solar cells, optoelectronic-devices, synthesis of thin layers and nanostructures including Quantum wells, quantum dots and nanowires. Dr. Das has also experience in Molecular-beam epitaxy (MBE), Plasma-enhanced chemical vapor deposition (PECVD), sputtering, Atomic Force Microscopy (AFM), Scanning Electron Microscopy (SEM), X-ray Diffraction (XRD) and Spectroscopy, UV and IR photoluminescence, transmission spectroscopy, time-resolved photoluminescence, Raman Spectroscopy, Hall Effect measurements, Spin coating, Reactive-ion etching (RIE), Electron beam evaporator, Rapid thermal processing (RTP), Electron-beam lithography, Photolithography, Wire bonding, Profilometer and Wafer saw, Optical Fibers (cleaving, cutting, splicing, launching light, loss measurements and characterization of various parameters), and in the operation of various Lasers, light-emitting diodes (LEDs), spectrophotometers, fluorometers, interferometers and photomultiplier tubes.

Biography of Professor Bimal Krishna Banik

Bimal Krishna Banik earned his Ph.D. degree in Chemistry at the Indian Association for the Cultivation of Science, Calcutta. Then, he pursued postdoctoral research at Case Western Reserve University, Ohio and Stevens Institute of Technology, New Jersey. Dr. Banik was a Tenured Full Professor and First President's Endowed Professor in Science & Engineering at the University of Texas-Pan American. He was the Vice President of Research & Education Development at the Community Health Systems of Texas. At present, he is Full Professor at the Prince Mohammad Bin Fahd University, Kingdom of Saudi Arabia.

Professor Banik has been conducting synthetic chemistry and medicinal research on cancers, antibiotics, natural products, and catalysis. As the Principal Investigator, he has been awarded $7.25 million grants from US NIH, US NCI, and US Private Foundations. He has more than 700 publications including 24 books, chapters, reviews, and papers, and more than 10,000 citations. In addition, he has 530 presentation abstracts.

Professor Banik has chaired 20 symposiums at the American Chemical Society National Meetings and over 2 dozen conferences at the National and International level, including 1 at the Nobel Prize Celebration in Germany. He has introduced approximately 300 speakers. He is a reviewer of 93, editorial board member of 28, editor-in-chief of 12, founder of 8, associate editor of 9, and guest editor of 10 research journals. He has recruited approximately 200 associate editors, regional editors, guest editors, and editorial board members for his journals. He is a reviewer of NSF, NCI, NRC, ACS, US Research Corporation, and International grant applications. He has evaluated approximately 2500 research grants and manuscripts. Professor Banik has served as Chair of the M.D. Anderson Cancer Center's drug discovery symposiums and directed their analytical chemistry laboratory. He has acted as a Chair/Member of more than 100 committees in his career.

Professor Banik has been involved in teaching science, medical, and engineering students. He has taught more than 3000 students in the USA and Saudi Arabia. He has established two science societies that have 1400 students at the University of Texas. Dr. Banik has mentored 20 postdoctoral fellows, 8 research scientists, and approximately 300 students. Importantly, he has mentored 28 faculty members in his research.

Dr. Banik received the Indian Chemical Society's Life-Time Achievement Award; Professor Asima Chatterjee Memorial Award; Mahatma Gandhi Pravasi Honor Medal from the UK Parliament; US National Society of Collegiate Scholars' Best Advisor Award for students; Professor P. K. Bose Endowment Medal; Dr. M. N. Ghosh Gold Medal; University of Texas Board of Regents' Outstanding Teaching Award; ACS Member Service Award; and Best Researcher, Teacher, and Mentor Awards by the UTPA. He is recognized by Burdwan University as an Eminent Alumni and also by ACS News, Elsevier, RSC, US NCI, US Research Foundations, University of Texas Board of Regents, Times of India, Stevens Institute of Technology, ACS SEED, NCI, NIH, India and US Newspapers, and Down to the Earth Magazine. Dr. Banik has received 200 invitations to deliver lectures in 50 countries. He is also invited to write research books by most reputed publishers in the world.

Preface

This book has four chapters targeting different photochemical research. The authors, Das and Banik, have written these diverse chapters using the available literature.

In Chapter 1, the use of photo-induced reactions for the synthesis of diverse organic compounds is described. It has been shown that light-mediated transformations are characterized by the involvement of electronically excited states that are created when photons are absorbed. A chemical compound participates this excitation method which results in the production of reactive intermediates and significantly alters the pathway of the compound. This can be controlled so that the intended natural product is produced in high yields and with excellent selectivity.

In Chapter 2, solar-activated photocatalysts is used as environmentally friendly and efficient in chemical processes. Photocatalysis allows initiation of conversions through novel mechanistic approaches. Some crucial examples of solar light-induced synthesis of biologically active compounds are described.

In Chapter 3, an environmentally friendly, photocatalytic method in the synthesis of compounds is demonstrated. This technology has gained a great deal of attention due to its performance in degrading toxic compounds under solar irradiation. Several organic reactions are catalyzed by different types of LED light sources (UV LED, blue LED, and white LED) and photocatalysts (transition metal complexes and organic dyes). The study explores the rate and yield of a reaction produced by LED lights, household bulbs, compact fluorescent lamp (CFLs), and halogen lamps. It discusses photo-induced electron transfers as well as energy transfer processes.

The use of photocatalysts against cancer is described in Chapter 4. Using photocatalysts to treat cancer is a new technique of photodynamic therapy. Three main components are involved in the PDT method, and they are oxygen, light, and photosensitizers. Hybrid semiconductor photocatalysts absorb light energy, then transfer it to oxygen, and generate cytotoxic reactive oxygen. Furthermore, nanophotocatalysts are capable of passing through biological barriers. Researchers have conducted studies in the hopes of achieving the goal of drug-free therapy. The recent advances in photo-mediated drug-free sustainable therapeutic approaches for the treatment of cancer are discussed.

The authors will be pleased if this focus book activates the research and education lives of students, scientists, and academic professionals along with creating business potential. Information provided in this book will also serve the curiosity of general people about photochemical processes.

This book will not be published without the assistance of CRC, Taylor & Francis Publisher. In particular, the authors are tremendously grateful to Ms. Hilary Lafoe and Ms. Varalika Kathuria for their knowledge in science, helping attitude, and encouragement.

Aparna Das
Bimal Krishna Banik

Photochemical Synthesis of Natural Products

1

Aparna Das and Bimal Krishna Banik

1.1 INTRODUCTION

Natural product synthesis represents synthetic chemistry's frontier, since it requires the practitioner to create molecular frameworks with complex and structurally distinct structures. Taking advantage of the wealth of synthetic challenges has provided a valuable platform for expanding state-of-the-art synthetic methodologies as well as identifying fundamentally new chemical protocols that can subsequently be implemented by the entire chemical community[1–3].

Drug researchers find that small-molecule libraries instilled with bioactive natural products are one of the most important aspects of the discovery process at the early stages of drug development. In order to develop new drug leads, natural products with biological activity are frequently discovered[4–6]. Chemists face a multifaceted challenge when attempting to synthesize these complex and structurally diverse assemblies, requiring the development of efficient and powerful synthetic strategies[7–9].

Simple feedstocks should be capable of being converted into highly functionalized and complex molecules if there were methods for doing so. A good idea would be to explore the potential of photochemical reactions, as they are limited to absorbing photons, which is a very attractive technique. In terms of their synthesis, photoinduced reactions are undoubtedly powerful and efficient ways of designing diverse organic frameworks that would otherwise be

DOI: 10.1201/9781003634249-1

difficult to construct[10,11]. As environmental issues have become more prominent in recent years, several existing technologies have also been reassessed, requiring the scientific community to devise "green" methods to address them. Besides being energy-efficient, these processes should also reduce raw material consumption, and ultimately, produce the least amount of waste[12,13].

In all light-promoted transformations, electronic excited states are formed when photons are absorbed. When a chemical compound undergoes this excitation process, transiently reactive intermediates are produced and the compound's reactivity is significantly altered, and this process can be controlled so that the intended product can be produced in high yields and with excellent selectivity[14,15].

It is possible to produce strained and unique targets molecules using the selective input of energy provided by light, which is not possible by conventional thermal protocols. By utilizing this method, it is now possible to produce enormous molecular complexity within a single chemical step. Most prevailing photochemical reactions are not enhanced by additional reagents, such as metal catalysts, acids, or bases, as opposed to thermal reactions. Therefore, rational and efficient synthetic methodologies can provide rapid access to diverse molecular scaffolds with a variety of functional groups, often in shorter synthetic sequences than alternative multistep synthetic techniques[16,17].

Photochemistry is especially attractive when it comes to generating molecular complexity that cannot be achieved with conventional methods. The application of light as a powerful tool for the construction of advanced polycyclic carbon skeletons has therefore been used to achieve a number of fascinating total synthesis of natural products[18,19]. Photocatalysis can also be conducted using artificial light sources, such as lamps, lasers[20], and light-emitting diodes (LEDs)[21]. Considering that we are involved in a variety of catalytic reactions[22-24], we found this topic very useful. We have prepared numerous beta-lactams by cycloaddition reactions using thermal, photochemical, and microwave-induced irradiation methods[25-36]. As part of our research, our group uses microwave-induced energy to prepare a variety of organic structures[37-55]. A key purpose of this chapter is to demonstrate the importance of photochemical approaches in accessing complex chemotypes and in synthesizing advanced structures that are relevant to biological systems.

1.2 SYNTHESIS OF BIYOUYANAGIN A

Biyouyanagin A was synthesized using a [2+2] photocycloaddition reaction, which required two raw material precursors, diene 3a (ent-7-epizingiberene) or 3b (ent-zingiberene) and enone 4[56]. As a result of irradiating a mixture of

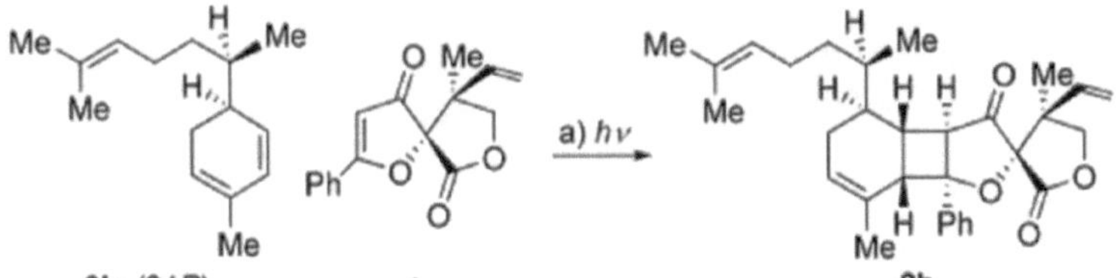

FIGURE 1.1   The total synthesis of biyouyanagin A (2b). (Adapted with permission from ref.[57].)

4 and 3a (24S) or 3b (24R) with 2′-acetonapthone as a triplet sensitizer and using a quartz cell (>320-nm filters) as a reaction vessel, it was able to achieve rapid chemo-, regio-, and stereoselective cyclobutane synthesis through hetero coupling, resulting in the formation of 2a (24S) or 2b (24R), respectively. In order to carry out the reaction, a fourfold excess of the less valuable terpene was added to a concentrated CH_2Cl_2 solution (**Figure 1.1**).

Biyouyanagin A (1, **Figure 1.1**) is a naturally occurring substance extracted from the leaves of *H. chinense L.* var. *salicifolium*, a species of *Hypericum*. This has been employed as a folk medicine for centuries for the treatment of female disorders in Japan[58]. Compared with noninfected H9 lymphocytes (EC_{50} = 0.798 μg mL^{-1}), this compound exhibited significant and selective inhibitory activity against HIV replication in H9 lymphocytes resulting in a therapeutic index of 31.3 in comparison to noninfected H9 lymphocytes (EC_{50} > 25 μgmL^{-1}). Moreover, biyouyanagin A in a strong manner suppressed lipopolysaccharide (LPS)-induced cytokine yield.

Several studies have been published on biyouyanagin A, including a total synthesis and structural revision[57]. Presented in the report is the total synthesis of both epimers of biyouyanagin A (24S and 24R) in their enantiomerically pure forms as well as the structural analysis of the natural product, which required not only the assigning of the R configuration to the C24 position but also the revision of the stereochemistry at the C17 and C18 positions (2a and 2b, **Figure 1.2**).

1.3 BRASOSIDE AND LITTORALISONE

On the formulation of (-)-brasoside, Iridoid 11 has been aggregated with 1-O-(TMS)-β-D-glucose tetraacetate, which upon deacetylation, in 13 overall steps. Using the intramolecular [2+2] photocycloaddition method, littoralisone was synthesized (**Figure 1.3**). After exposure to ultraviolet (UV) light

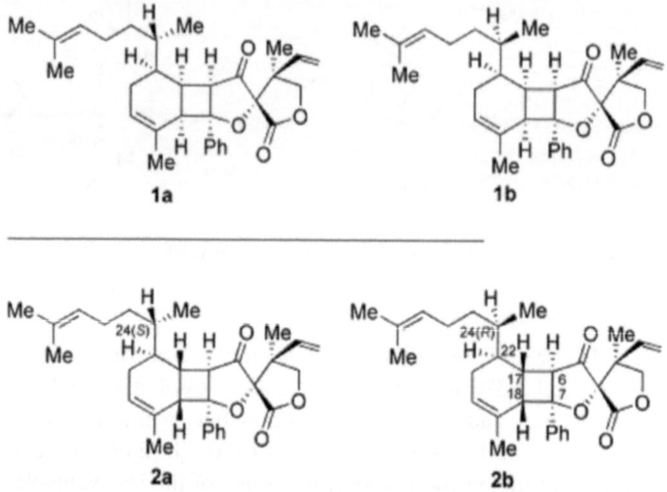

FIGURE 1.2 Originally proposed (1a and 1b) structures of biyouyanagin A and revised (2a and 2b) structures. (Adapted with permission from ref.[57]).

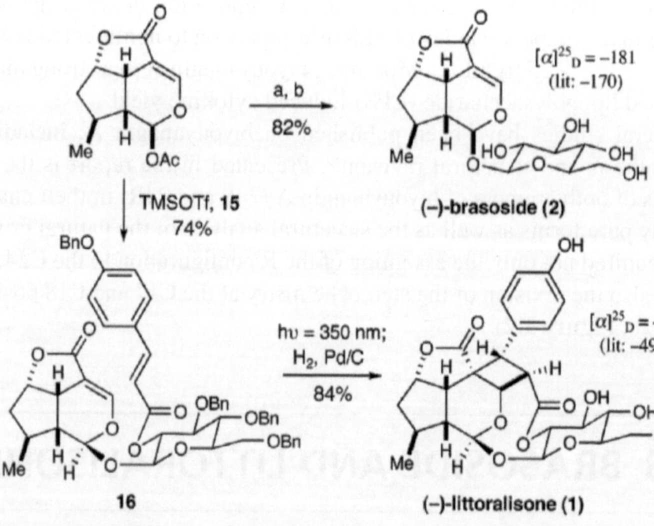

FIGURE 1.3 Synthesis of brasoside and littoralisone. (Adapted with permission from ref.[59]. Copyright {2005} American Chemical Society.)

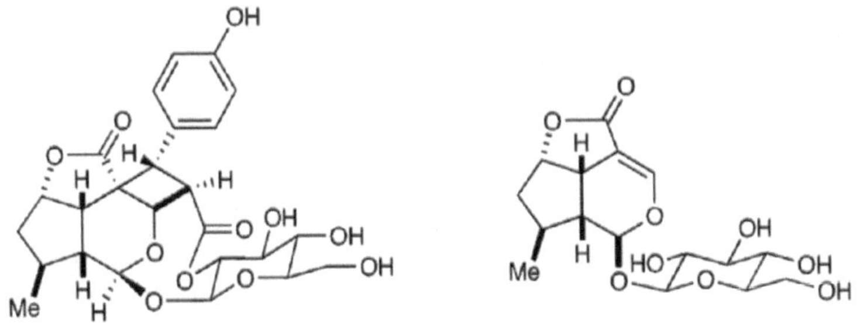

(−)-littoralisone (1)   (−)-brasoside (2)

FIGURE 1.4   Molecular structure of brasoside and littoralisone. (Adapted with permission from ref. [59]. Copyright {2005} American Chemical Society.)

(350 nm) for 2 hours and hydrogenolysis in situ, 16 yielded (-)-littoralisone with a yield of 84%, similar to the natural isolate. In ambient light, it was also observed that [2+2] cycloaddition occurred slowly, suggesting that the proposed biochemical process of littoralisone formation may be plausible.

In the literature, a complete synthesis was reported for brasoside and littoralisone (**Figure 1.4**)[59]. A total of 13 steps were required to synthesize littoralisone (1) and brasoside (2). In order to make both natural products, an intermediate is quickly assembled using organocatalytic technology, including a proline-catalyzed α-aminoxylation and a contra-thermodynamic intramolecular Michael addition. Utilizing a two-step carbohydrate synthesis technology, glucose derivatives were synthesized with an intramolecular cycloaddition tether that was selectively substituted for intramolecular cycloadditions. Interestingly, this synthesis culminates in the intramolecular [2+2] photocycloaddition, which supports the biosynthetic origin of 1 from 2. Over the past few years, we have made significant progress in various catalytic procedures. Photo radiation can also have a powerful impact on synthesis[22,23,60–64].

1.4 SYNTHESIS OF PUNCTAPORONIN C

The punctaporonin C (**Figure 1.5**) was isolated along with diverse structurally similar compounds, the punctaporonins[65,66], from *Poronia punctata*. It belongs to the caryophyllene family and belongs to the sesquiterpene. The tetracyclic sesquiterpene Punctaporonin C was synthesized[67].

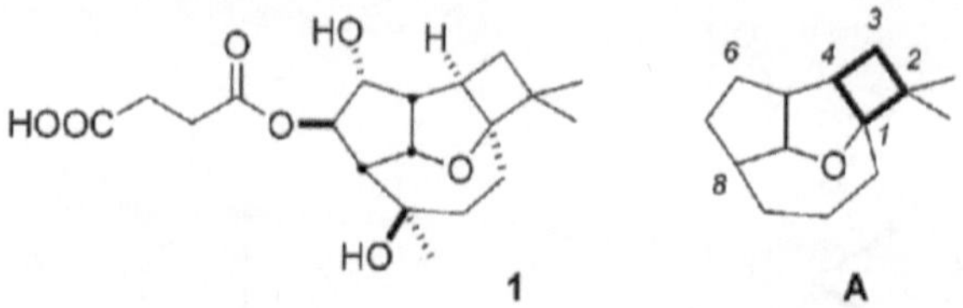

FIGURE 1.5 Structure of punctaporonin C (1) and its core skeleton A with the cyclobutane ring highlighted. (Adapted with permission from ref. [67].)

It has been discovered, however, that 1,3-divinyl-2-cyclopentyltetronates with a polar substituent at position 4 undergo cycloaddition in protic solvents, giving predominantly the expected regioisomer. Therefore, it was found that compound 3 was obtained very efficiently from compound 2 with excellent selectivity (**Figure 1.6**). Based on the hypothesis, it is proposed that the acetoxy group of envelope 2′ becomes sterically demanding as a result of hydrogen bonding to the residues and solvent in a pseudoequatorial region. Due to this, the tetronate group and one of the two terminal double bonds of the molecule are in the perfect location for an intramolecular [2+2] photocycloaddition reaction. Using a photoreaction with perfect facial and simple diastereoselectivity, a single diastereoisomer was obtained.

The common meso epoxide 4 was used as the starting material for the successful production of punctaporonin C (Figure 1.7). By opening the epoxide with KOAc, the free hydroxy group is protected as its TIPS (triisopropylsilyl) ether. Through cleavage of TBDMS ether, alcohol 5 was produced, which was then converted into tetronate 2. The use of chloromethanesulfonate as a

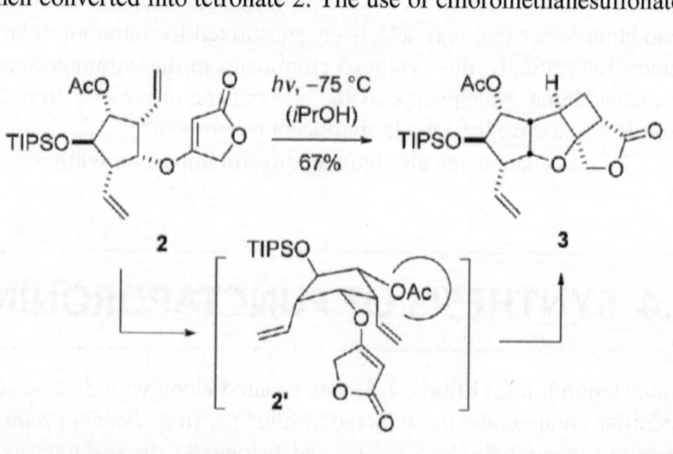

FIGURE 1.6 Regio- and stereoselective [2+2] photocycloaddition. (Adapted with permission from ref. [67].)

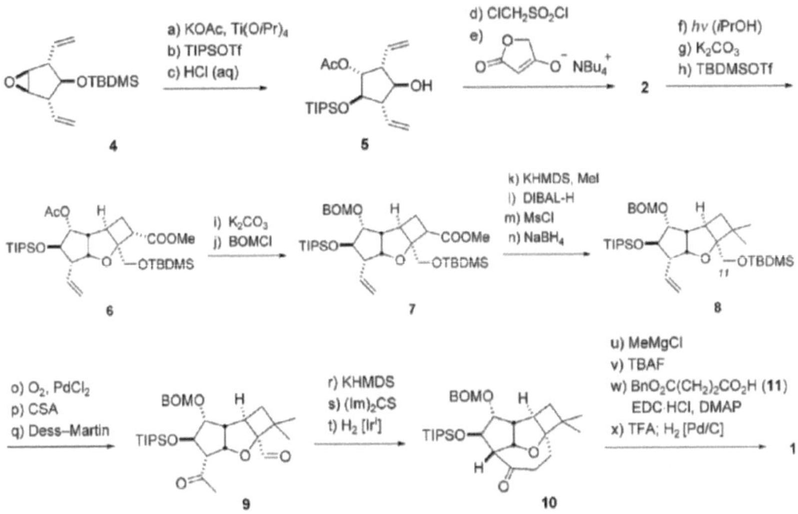

FIGURE 1.7 Complete synthesis of punctaporonin C. (Adapted with permission from ref. [67].)

leaving group was important for the high-yielding formation of product 2 in the alkylation process induced by configuration inversion. Instabilities were observed with trifluoromethanesulfonate at the same time as a maximum yield of 30% with methanesulfonate.

Lactone opening was observed under mild conditions. Due to the intramolecular [2+2] photocycloaddition of 2, its free primary alcohol was also protected, along with its TBDMS ether. An acetyl group was placed by a BOM (benzyloxymethyl) group in compound 6 because the acetyl group was unable to protect the alkyl group during the planned alkylation. The cyclobutane ring, however, underwent further transformation in ester 7. Upon introducing the methyl group via the enolate alkylation process, the second methyl group is formed upon complete reduction of the exocyclic methoxycarbonyl group.

As the Wacker oxidation of compound 8 was efficient and yielded a high product yield of 94%, the intramolecular production of the oxepane ring was examined by enolate alkylation after primary alcohol deprotection and the installation of appropriate leaving groups at C11. As a result of the adjacent geminal dimethyl substitution, this transformation was unable to succeed, probably because the leaving group could not locate itself properly in an SN$_2$-type reaction.

Aldol reactions enabled the closure of the ring by oxidizing the free alcohol to the aldehyde 9. In a chemoselective manner, the double bond was eliminated and hydrogenated to obtain the desired ketone 10. Due to a pseudoaxial reaction at low temperatures, a Grignard addition with readily accessible

MeMgI resulted in the formation of the wrong diastereoisomer. By varying the counterion and proceeding with the reaction at RT, the corresponding tertiary alcohol was obtained with very good diastereoselectivity.

The final three steps of the process went smoothly. In order to access the target molecule 1, the succinic acid side chain was added as its monoprotected derivative 11. Neither hydrogenolysis of the BOM protecting group nor hydrogenolysis of the benzyl ester was successful. Therefore, under acidic conditions, the BOM group must be removed first, followed by hydrogenolysis to cleave the benzyl ester.

1.5 ISOLATION AND PREPARATION OF PHYTOALEXINS

Solavetivone 1 inhibits mycelial and germ tube growth, germination, and important enzymes of Pseudomonas infestans. Furthermore, it is also capable of inhibiting the growth of Pseudomonas syringae pv. tabaci. Also, 3-hydroxysolavetivone, solavetivone 1, and solanascone 3 have strong antibacterial activity against *Pseudomonas solanacearum*. Enantiospecific total synthesis of phytoalexins, (+)-solanascone, (+)-dehydrosolanascone, and (+)-anhydro-β-rotunol was reported[68].

In order to protect themselves from pathogens, plants produce phytoalexins, which are low-molecular-weight secondary metabolites. As with antiviral proteins, phytoalexins inhibit protein synthesis during a pathogenic attack, preventing the foreign agent's growth by removing any factors that could allow the pathogen to invade. The phytoalexins anhydro-rotunol 2 and solavetivone 1 have been found to be important stress metabolites in potato tubers affected by soft-rot bacteria (such as *Erwinia carotovora var atroseptica*) or blight fungi (such as *Phytophthora infestans*). Solavetivone 1 has also been isolated from potato cell suspension cultures and air-cured tobacco leaves.

It was from Nicotiana tabacum cv. Burley that the structurally complex phytoalexin (+)-solanascone 3 was first isolated. The presence of dehydrosolanascone 4 in flue-cured tobacco leaves was discovered several years later. Following this, a few other members of the solanascane family were extracted (**Figure 1.8**).

Solanascone 3 can be synthesized via a retrosynthetic process shown in **Figure 1.9**. Enone 5 was identified as an important intermediate in the synthesis of solanascone 4 and solanascone 3, as well as anhydro-rotunol 2 and solavetivone 1. To obtain enone 5 from the dione 6, an intramolecular aldol reaction can be used. A suitable precursor for dione 6 is diketone 7, in which the but-3-enyl group can be viewed as the masked 3-oxobutyl group.

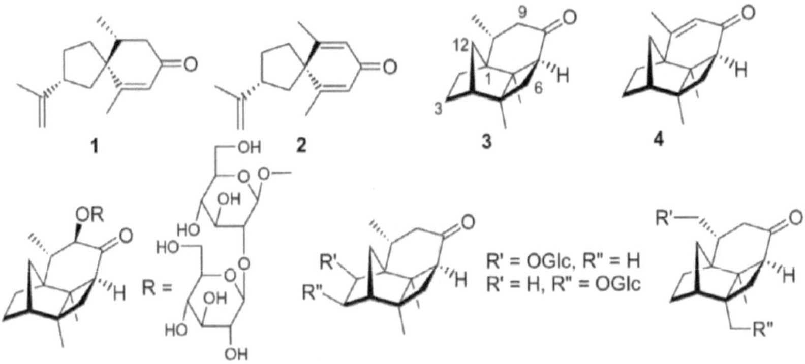

FIGURE 1.8 Examples of phytoalexins. (Adapted with permission from ref. [68].)

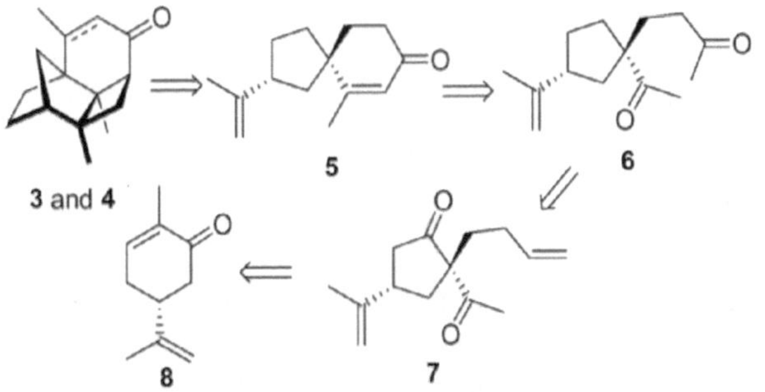

FIGURE 1.9 Retrosynthetic analysis of solanascone 3. (Adapted with permission from ref. [68].)

3-Butenylcarvone could be used to create compound 7 (2-acetyl-2-butenylcyclopentanone) from compound 8 (carvone).

In addition, it was also investigated whether dehydrosolanascone 4 and solanascone 3 could be synthesized (**Figure 1.10**). Due to the insertion of the C-10 methyl group in solavetivone 1, an epimeric mixture was formed (more clearly 16 to 1+17). The structure of the tetracyclic carbon framework was used to direct the insertion of the secondary methyl group to the exo face after the construction of the tetracyclic carbon framework. Under controlled conditions, the resulting tetracyclic ketone 18 was produced by photochemically irradiating a degassed methanolic solution of enone 5 with a 450-W mercury vapor lamp for 2 hours.

The TMS enol ether of ketone 18 was prepared by treating it with trimethylsilyl chloride, LHMDS, and triethylamine in THF at −70°C, followed by treatment with palladium acetate in acetonitrile at room temperature to

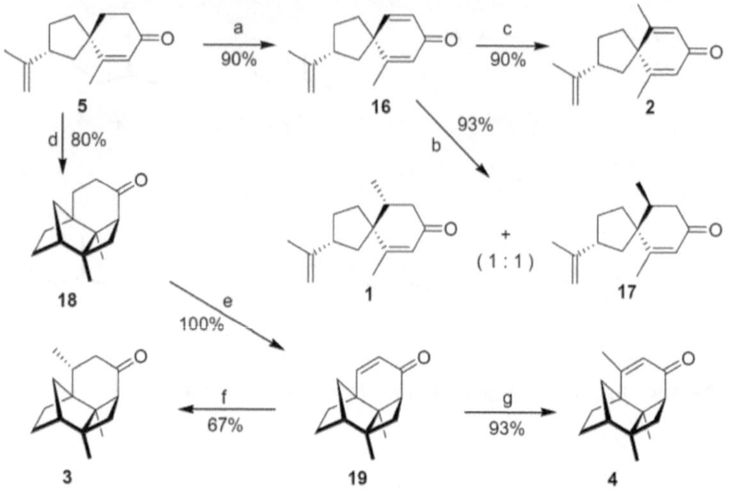

FIGURE 1.10   Syntheses of dehydrosolanascone 4 and solanascone 3. (Adapted with permission from ref. [68].)

give the enone 19. The compound 3 was obtained in a 67% yield by treating dehydro-nor-solanascone 19 with lithium dimethylcuprate at −70°C and followed by stirring at −30°C for 1.5 hours. A 93% yield of dehydrosolanascone 4 was obtained by reacting enone 19 with trimethylsilyl chloride, lithium dimethylcuprate, and triethylamine, followed by treating the TMS enol ether of this reaction with palladium acetate in acetonitrile.

1.6   ISOLATION AND SYNTHESIS OF HUMULANOLIDES

Recently, the community has paid considerable attention to the sesquiterpenes in Asteriscus (Compositae). (-)-Asteriscunolide A [(-)-1], which was isolated from *Asteriscus aquaticus* together with its cis-trans isomers (-)-Asteriscunolides B—D [(-)-2—4], represents a sesquiterpene lactone containing a humulane skeleton, that is known as a humulanolide (**Figure 1.11**). Additionally, the plant contains a tricyclic (+)-asteriscanolide [(+)-5]. The same plant has also been found to yield tricyclic (+)-asteriscanolide [(+)-5], as well as its tetradehydro analogue (+)-tetradehydroasteriscanolide [(+)-6] from A. graveolens. In 1989, San Feliciano discovered humulanolides, called aquatolides, that had never been seen before in *A. aquaticus*.

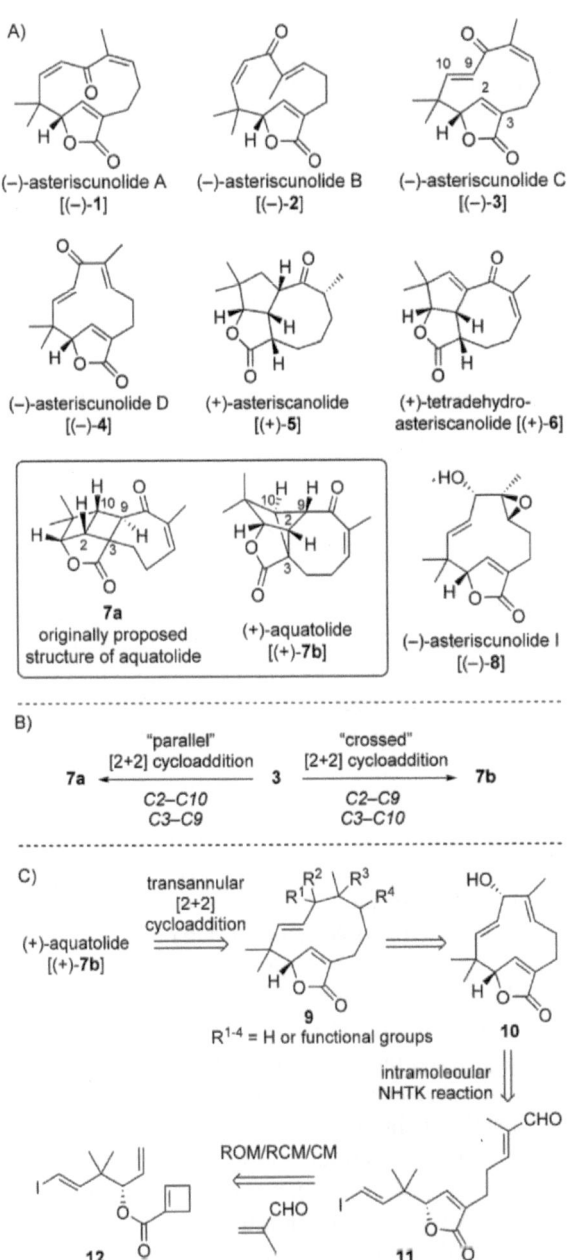

FIGURE 1.11 (A) Aquatolide and related humulanolides. (B) Biosynthesis of aquatolide. (C) Retrosynthetic analysis of aquatolide. (Adapted with permission from ref. [69].)

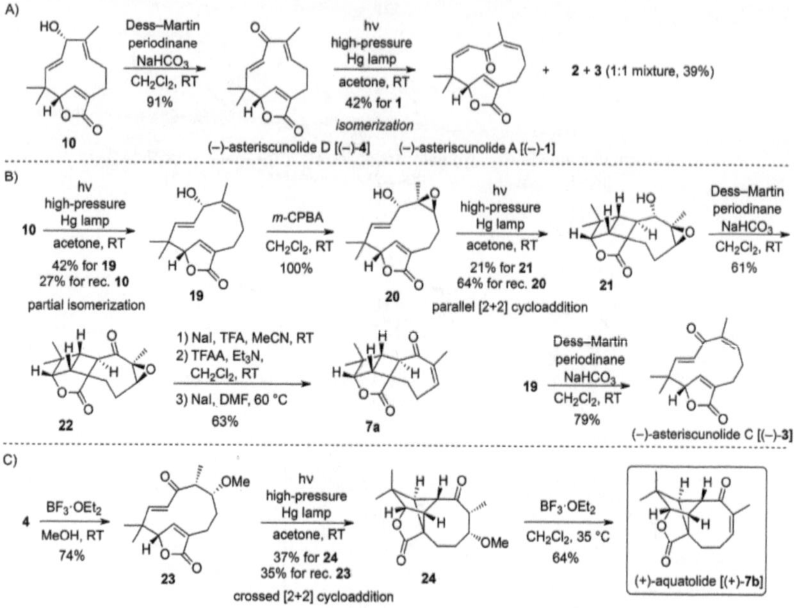

FIGURE 1.12 (A) Syntheses of asteriscunolides. (B) Syntheses of the originally proposed structure of aquatolide (7a) and asteriscunolide C [(-)-3]. (C) Total synthesis of aquatolide [(+)-7b]. (Adapted with permission from ref. [69].)

First, the oxidation of 10 to (-)-4 was performed (**Figure 1.12A**). Through Pyrex glass, the Hg lamp, emitting a wide range of wavelengths, was used to irradiate (-)-4. In this experiment, the results were similar to those obtained previously with a UV lamp (10 W, 254 nm), and only the isomerization of the olefins took place. As a result of the process, (-)-1 was isolated in a yield of 42% (a total yield of 14% in 11 steps).

After irradiating 10, instead of (-)-4, the trisubstituted olefin was chemoselectively isomerized into a Z configuration, yielding 19 with a yield of 42%. However, the subsequent [2+2] cycloaddition failed (**Figure 1.12B**). It was suspected that the olefin would need to be masked before it could be used. In order to achieve the desired results, 19 was treated with m-CPBA. The single isomer 20 was produced by chemo- and stereoselective epoxidation. As expected, the [2+2] photocycloaddition of 20 went smoothly. Nevertheless, the cycloaddition mode of the reaction was parallel and 20 was converted into the [2]-ladderane adduct 21 in a yield of 21% (58% brsm). By oxidizing 21 and deoxygenating the resulting epoxide 22 by iodohydrin, aquatolide (7a) was formed (originally proposed structure).

Aquatolide's total synthesis was also reported in another study[69]. Note that the main reaction step consists of the intramolecular [2 + 2] photocycloaddition of an allene to an α, β-unsaturated δ-lactone. Additionally, a Horner–Wadsworth–Emmons reaction is required to close the lactone, followed by a Mukaiyama-type aldol reaction to cyclize the eight-membered ring intramolecularly. The racemic aquatolide has been resolved using preparative HPLC.

1.7 ISOLATION AND SYNTHESIS OF SOLANOECLEPIN A

A complex naturally occurring terpenoid, solanoeclepin A (1), is produced by the roots of young potatoes in spring (**Figure 1.13**). This compound has been shown to have nanomolar activity as a hatching agent for potato cyst nematodes (PCN). The presence of these parasitic worms (maximum length 1 mm), which feed on potato roots, has resulted in considerable losses in potato harvests in many countries. As a result of the PCN entering the root system of the potato after hatching (from cysts in the soil), the potato's growth is retarded.

The formal synthesis of solanoeclepin A has been described in a report[70]. The tricyclic core was constructed by enantioselective allene diboration and intramolecular [2+2] photocycloaddition. An enantioselective synthesis of an intermediate in the Tanino total synthesis of solanoeclepin A has been developed. In six different steps, the tricyclo[5.2.1.01,6]decane core was obtained by an intramolecular [2+2] photocycloaddition. For the preparation of the first photosubstrate, an indium-mediated Barbier reaction was used, which produced an excellent [2+2] cycloaddition, but the enantiopurity wasn't sufficient. When irradiated at 365 nm on a 20-g scale in a flow system, the [2+2] cycloaddition product was obtained in a high yield using the second

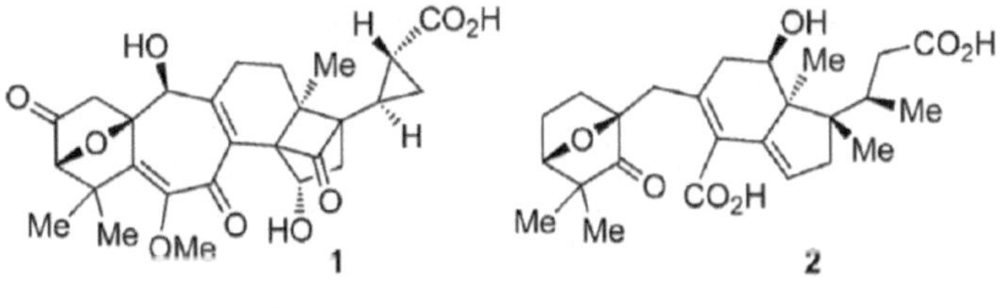

FIGURE 1.13 Structures of glycinoeclepin A (2) and solanoeclepin A (1). (Adapted with permission from ref. [70].)

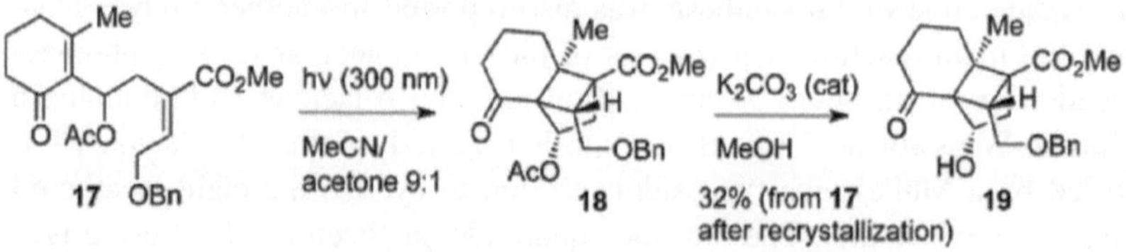

FIGURE 1.14   Photocycloaddition of the ester 17. (Adapted with permission from ref. [70].)

photosubstrate, which was prepared through an asymmetric allene diborylation in high enantiomeric excess. Also, important steps included diastereoselective cyclopropanation of allylic alcohol and replacing a boronate group with a vinyl group at the quaternary carbon.

Acetate 17 was irradiated with 300-nm UV-light in a 9:1 mixture of acetonitrile and acetone in a Rayonet photoreactor to perform the key photocycloaddition (**Figure 1.14**). The starting material was converted within 2 hours into four inseparable isomeric products in a ratio of approximately 7:1:1:1, according to 1H NMR. From this mixture, the primary product was the desired 18 since after removing the acetyl group and recrystallizing, pure racemic alcohol 19 was produced in a 32% yield.

1.8 SYNTHESIS OF CYCLOBUTENES

By using an organic single electron oxidant to electron relay (ER) system, cyclobutene lignans can be synthesized[71]. A photoinduced electron transfer method can be used to synthesize lignan cyclobutanes and their analogs directly. Many oxygenated alkenes can provide terminal or substituted cyclobutane adducts with complete stereochemical control, resulting in trans-stereochemical cycloadducts. To minimize competing cycloreversion, it is necessary to include an aromatic ER. Natural products magnosalin and pellucidin A have been synthesized using this method.

[2+2] Cycloaddition of 1,3-Dienes has been reported via visible light photocatalysis[72]. In synthetic chemistry, 1,3-diene's photocycloadditions present a useful yet underutilized group of reactions. Using the [2+2] cycloaddition process, 1,3-dienes could be cycloadditioned with photo-absorbing transition metal complexes. Because long-wavelength visible light is used for direct photoexcitation of 1,3-dienes, this reaction method is attractive because under such reaction conditions, there is no risk of decomposition of sensitive functional groups that would otherwise be easily decomposed by Ultraviolet C (UVC) radiation during direct photoexcitation. It is possible to use the obtained vinylcyclobutane for a variety of further synthesis reactions. In the near future, this method should be able to facilitate the production of complex organic targets. **Figure 1.15** shows an overview of experiments probing the effects of substrate modifications on the [2+2] cycloaddition.

In photochemical reactions based on visible light, transition metal photocatalysts have been widely used for photoinduced redox activation of organic functional groups. Through an energy-transfer process, it is possible to conduct a [2+2] cycloaddition of various electronically diverse styrenes using iridium complex 2. Visible light is also expected to be capable of initiating [2+2] cycloaddition reactions via diene activation. While dienes are more resistant to one-electron oxidation than styrenes, their lowest lying triplet states have a similar energy level. As a result, it is likely that the same visible light-activated catalysts that were effective in sensitizing styrenes might also provide a broad range of synthetically important vinylcyclobutane derivatives (**Figure 1.16**).

A study of the sensitization of higher-order conjugated polyenes was also conducted (**Figure 1.17**). In the presence of 2·PF6, upon irradiation, the compound 20 underwent high-yielding cycloaddition to compound 21. Ultimately, the compound was formed as a 1:1 mixture of E and Z isomers.

Due to the fact that photosensitization reactions use low-energy, operationally convenient, and readily available visible light as opposed to UV irradiation, these processes are appealing for synthetic applications. To illustrate this point, the reaction of vinyl iodide 32 under optimal conditions was compared to the reaction under direct UV irradiation (**Figure 1.18**). A Rayonet reactor equipped with 254-nm lamps was used to conduct the experiments, and it was found that 32 is rapidly consumed. It is found that the mass recovery of the reaction is poor, which is consistent with high-energy UV radiation's propensity to promote uncontrolled radical decomposition processes. In fact, cyclobutane 23 decomposed completely within 1 hour of being re-irradiated with UV light at 254 nm. Therefore, these conditions are not only more convenient than traditional UV photolysis, but are also more tolerable to photochemical reactions with photosensitive functional groups, such as alkyl iodides.

In order to demonstrate that the conditions established above are applicable to the preparation of complex natural products, cannabiorcicyclolic acid (37) was synthesized, one of several cyclobutane-containing cannabinoids (**Figure 1.19**). Chromene precursor 35 can be obtained with a yield of 54% by adding base-catalyzed condensation between phenol 33 and citral (34). Photocycloaddition sensitized by 2·PF6 produced cyclobutane 36 in a high yield (86%) after 8 hours of irradiation with a 23-W compact fluorescent light

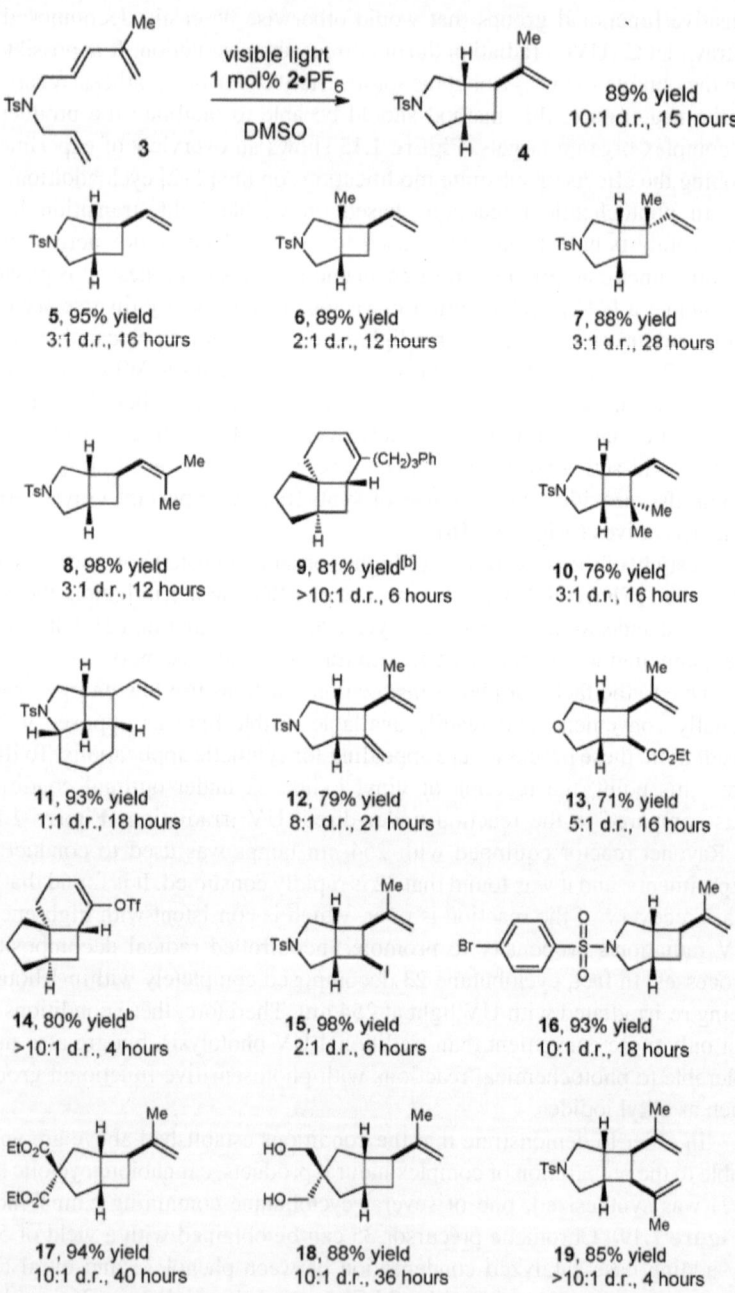

FIGURE 1.15   Various compounds. (Adapted with permission from ref. [72].)

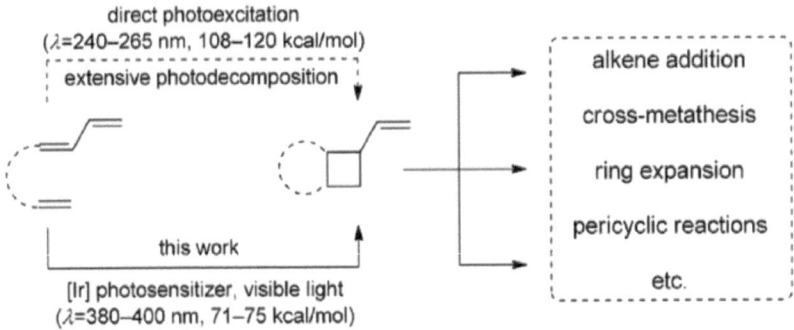

direct photoexcitation
(λ=240–265 nm, 108–120 kcal/mol)

extensive photodecomposition

this work

[Ir] photosensitizer, visible light
(λ=380–400 nm, 71–75 kcal/mol)

alkene addition

cross-metathesis

ring expansion

pericyclic reactions

etc.

FIGURE 1.16 [2+2] Cycloadditions of 1,3-dienes. (Adapted with permission from ref. [72].)

20

TsN ... Me Me

visible light
catalyst (1 mol%)
DMSO

TsN ... Me Me

21

Ir conditions: 1 mol% **2**•PF$_6$, 18 hours: 96% yield, 1:1 *E/Z*, 2:1 d.r.

Ru conditions: 1 mol% **1**•(PF$_6$)$_2$, 4 hours: 93% yield, >10:1 *E/Z*, 2:1 d.r.

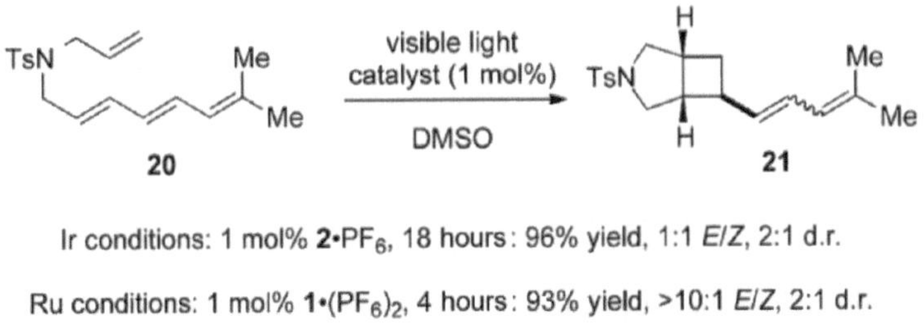

FIGURE 1.17 [2+2] Cycloaddition of higher order polyenes. (Adapted with permission from ref. [72].)

32

conditions

23

23 W CFL, **2**•PF$_6$
254 nm, no sensitizer

15 h, 90% yield
2 h, 0% yield

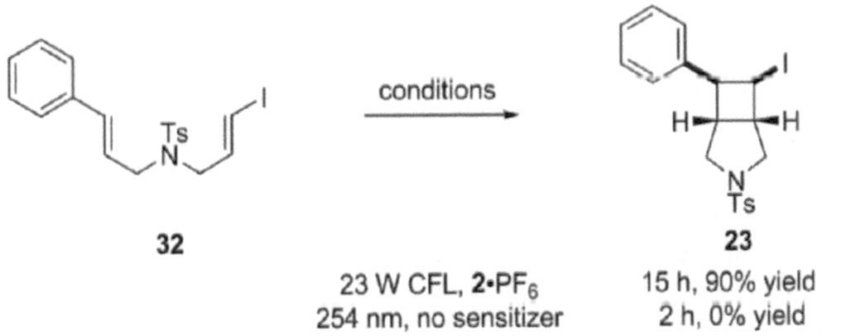

FIGURE 1.18 For cycloaddition of compound 32, comparison of visible to UV conditions. (Adapted with permission from ref. [73].)

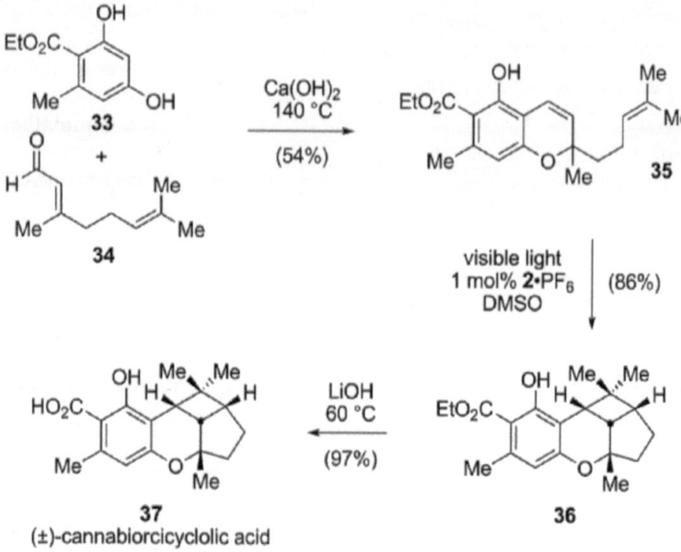

FIGURE 1.19   Synthesis of cannabiorcicycloic acid. (Adapted with permission from ref. [73].)

bulb. Directly irradiating 35 with 254-nm UV light for 5 hours produced only 19% of the desired cycloadduct, with only 9% of the starting chromene remaining. Cannabiorcicycloic acid can be hydrolyzed from ethyl ester in a 97% yield under typical ester hydrolysis conditions.

The visible light photocatalysis of the [2+2] cyclotransformation of styrene has been shown to be effective[73]. The prospect of conducting synthetically useful organic reactions with visible light has attracted the attention of several research groups in recent years. Polypyridyl complexes of ruthenium and iridium have been the focus of these efforts, and they have demonstrated that when exposed to visible light, these complexes can undergo a variety of photocatalytic transformations due to their remarkable photoredox properties.

As an alternative to UV light, visible light can be used in organic photochemistry to achieve a number of benefits. In addition to the lower costs and lower energy demands of visible light sources, photoreactions can also be conducted without special photoreactors or quartz glassware, as well as selectively photoexciting a transition metal photocatalyst without triggering unwanted radical reactions of organic functional groups that are photochemically sensitive. The exploration of special properties of metals led us to become interested in their uses. The metal tellurium is an example of one of these metals[74–85].

1.9 SYNTHESIS OF GLIOCLADIN C

To introduce triketopiperazine and the unsaturated imide in a single step, as part of the synthesis, enamine formation and intramolecular amidation were carried out under N-acyliminium ion promoter conditions (**Figure 1.20**; $25 \rightarrow 24$). Therefore, it was found that by sequentially treating the secondary amine with DBU and NBS, it was possible to obtain the requisite imine 25 with nearly quantitative yields as a result of the oxidation. As a result of the microwave irradiation of a mixture of ClCOCO$_2$Et, imine 25, and Et$_3$N at 150°C in toluene, the acylation/cyclization step was carried out. An acyliminium intermediate 26 is produced, as a result of the reaction of the imine with ClCOCO$_2$Et, which upon deprotonation forms an enamine 27. There is an intramolecular amidation reaction in which the amide nitrogen atom attacks the ester carbonyl as a means of closing the ring, thereby resulting in the production of triketopiperazine 24 in a 76% yield (after one recycle). It was concluded that global Cbz removal using BCl$_3$ in CH$_2$Cl$_2$ (78°C–23°C) resulted in gliocladin C (1) being obtained in a 80% yield in the following step.

By using visible light photoredox catalysis, a total synthesis of (+)-Gliocladin C has been achieved[86]. Natural products that fall under the hexahydropyrroloindoline alkaloid family include a large number of compounds derived formally from two molecules of tryptophan[87]. A subset of

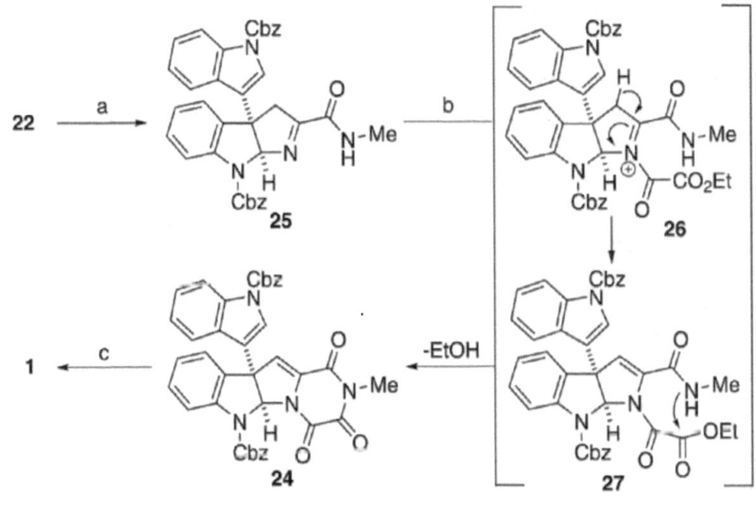

FIGURE 1.20 Synthesis. (Adapted with permission from ref. [86].)

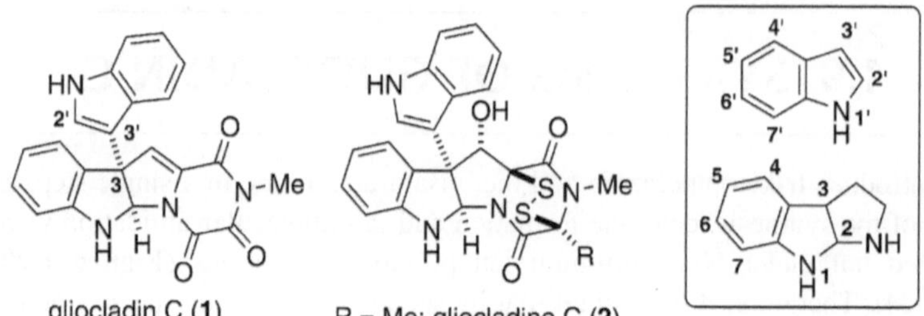

gliocladin C (1)

R = Me; gliocladine C (2)
R = H; bionectin A (3)
R = CH$_3$CH(OH); bionectin B (4)
R = iPr; leptosin D (5)

numbering scheme for
indole (top) and
pyrroloindoline
(bottom) moieties

FIGURE 1.21   Examples of antibiotic and cytotoxic C3–C3′ bisindole alkaloids. (Adapted with permission from ref. [86].)

this class of molecules, called the C3-C3′ indole alkaloids, have the same 3a-(3-indolyl)-hexahydropyrrolo-[2,3-b]indole skeleton. Compounds such as gliocladin C[88], gliocladine C[89], leptosin D[90], and bionectins[91] fall into this category (**Figure 1.21**). Furthermore, these compounds possess a range of potent biological properties in addition to their interesting structural features. Gliocladin C and leptosin D, for example, are cytotoxic against P-388 lymphocytic leukemia cell lines with ED50 values of 240 and 86 ng mL^{-1}, respectively. On the other hand, bionectins A and B have antibacterial activity against both methicillin-resistant *Staphylococcus aureus* (MRSA) and quinolone-resistant *S. aureus* (QRSA) with MICs of 10–30 µM mL^{-1}. Our research group has investigated a number of medicinally active compounds. Several new sugars, natural products, and carbocyclic and heterocyclic organic compounds have been synthesized as a result of these efforts[24,92–98].

The synthesis of 1 began with an orthogonal nitrogen protection of Boc-D-tryptophan methyl ester (17) using CbzCl as an initiator (**Figure 1.22**). Using NBS and PPTS, bromopyrroloindoline 18 was produced in a 91% yield by the two steps of bromocyclization[99]. A methylamidation of 18 in THF with aqueous MeNH$_2$ produced the corresponding methylamide 19 with a yield of 87%. Bromopyrroloindoline 19 was subjected to the key indole coupling reaction after optimization of the reaction conditions. As a result of blue-light irradiation with 1 mol% of [Ru(bpy)3Cl$_2$] in DMF, amide 19, and aldehyde 15 (5.0 equiv) provided an 85% yield of the desired coupling product 20, with Et$_3$N (2.0 equiv).

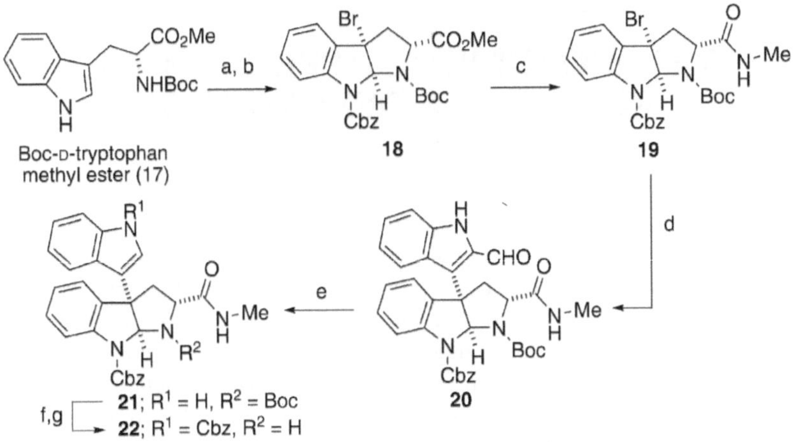

FIGURE 1.22   Synthesis. (Adapted with permission from ref. [86].)

1.10 SYNTHESIS OF PSEUDOTABERSONINE, PSEUDOVINCADIFFORMINE, AND CORONARIDINE

Preparation of (−)-pseudotabersonine, (−)-pseudovincadifformine, and (+)-coronaridine enabled by photoredox catalysis in flow was reported[100]. The use of photoredox catalysis for natural product modification has been demonstrated, which allows chemoselective access to a wide range of chemical structures in complex chemical space, providing both useful quantities of less abundant congeners and the possibility of new structural motifs. The controlled modification of amine substrates through single-electron oxidation is perfect for the synthesis and modification of alkaloids, even though amine additives have long been used as stoichiometric electron donors for photocatalysis. The conversion of the amine (+)-catharanthine into the natural products (−)-pseudotabersonine, (−)-pseudovincadif-formine, and (+)-coronaridine utilizing visible light photoredox catalysis was examined in detail.

1.11 BIOMIMETIC APPROACH TO ROCAGLATE DERIVATIVES

Biomimetic approaches to rocaglamides have been proposed using the photo-generation of oxidopyryliums derived from 3-hydroxyflavones (3-HF)[101]. The innovation resulted in a unified biomimetic approach for aglain, forbaglin, and rocaglamide classes of natural products. In this procedure, oxidopyryliums are generated by excited-state intramolecular proton transfer (ESIPT) from 3-HFs, followed by [3+2] dipolar cycloaddition to the aglain core. An acyloin rearrangement at the substitution site rearranges the aglain core into the rocaglamide framework. This approach was successfully employed for the synthesis of the natural product (±)-methyl rocaglate.

There are several secondary metabolites produced in the genus Aglaia, including cyclopenta[b]benzofurans[102], rocaglamide (1), silvestrol (2), and cyclopenta[b, c]benzopyrans[103], with ponapensin 3 being one of them (**Figure 1.23**). It is known that cyclopenta[b]-benzofurans are potent anticancer agents because they modulate the activity of eukaryotic initiation factor 4A (eIF4A), an RNA helicase involved in loading ribosomes onto mRNA templates when translation is initiated. It is often deregulated in cancer[104]. In light of its

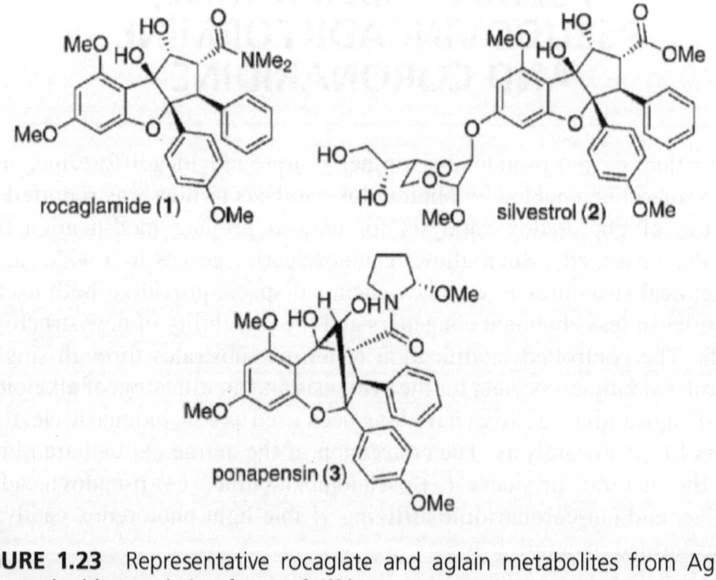

FIGURE 1.23 Representative rocaglate and aglain metabolites from Aglaia. (Adapted with permission from ref. [122].)

unique structure and biological activity, rocaglamide (1) has been targeted by a number of research groups and inspired a number of inventive synthetic methods[105]. We examined anticancer molecules[106–121] in our study and found that the method and compounds used in this investigation have similar functions.

An asymmetric synthesis of rocaglamides was achieved by enantioselective photocycloaddition with chiral Bronsted acids[123]. In addition to methyl rocaglate, rocaglamide and rocaglaol have been successfully synthesized by enantioselective procedures. One method used to obtain an aglain precursor followed by a ketol shift/reduction sequence to the rocaglate core is enantioselective [3 + 2] photocycloaddition via chiral Brønsted acids (TADDOLs).

There has been reported a biomimetic process to the cyclopenta[b] benzofuran natural compounds, involving a photocycloaddition/ketol shift rearrangement/reduction sequence involving 3-HF derivatives, such as compound 4 and compound 5a (methyl cinnamate)). Using ESIPT (excited-state intramolecular proton transfer) of 3-HFs as the proton transfer mechanism, both 1 and 2 were synthesized[122]. Upon photoirradiating 4, the oxidopyrylium intermediate 6 is formed, which undergoes a [3+2] photocycloaddition with 5a to form the aglain core 7a, which is then moved into the methyl rocaglate (8a) (**Figure 1.24**). In addition, the scope of the photocycloaddition, the photophysical and mechanistic studies, and the evaluation of the rocaglates as eukaryotic translation inhibitors were discussed.

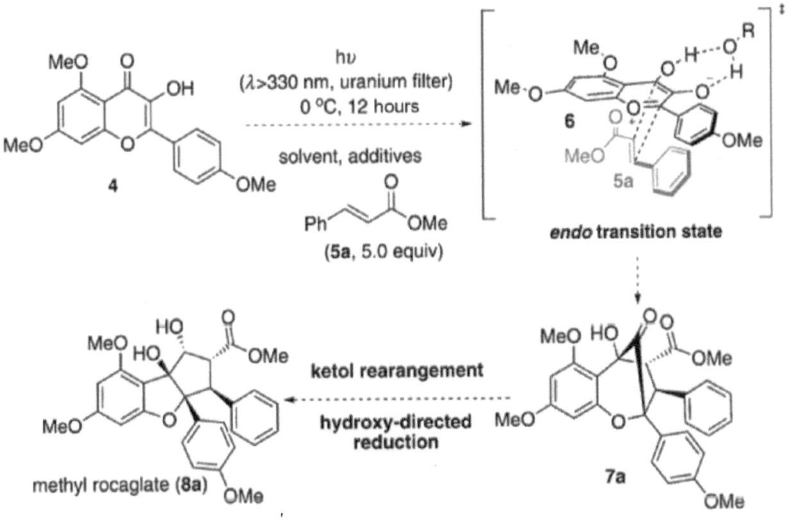

FIGURE 1.24 To the cyclopenta[b]benzofurans, the [3+2] photocycloaddition approach. (Adapted with permission from ref.[122].)

The biomimetic photocycloaddition of 3-HFs for the synthesis and evaluation of rocaglate derivatives as eukaryotic translation inhibitors was reported[124]. Several dipolarophiles (5) have been tested for their reactivity in the [3+2] photocycloaddition under optimal conditions. There were cinnamate derivatives modified at both termini (easily prepared or commercially available) as well as unactivated alkenes in the set of dipolarophiles evaluated. Therefore, a wide range of dipolarophiles could be used in photocycloaddition reactions.

It was observed, however, that all beta-alkylacrylate derivatives were unreactive under the reaction conditions (yields of less than 10%), which is most likely due to a lack of radical stabilization or a positive charge on the beta position of the unsaturated ester, both of which may have contributed to the unreactivity. The population of oxidopyrylium species 6/6' increases significantly with the use of TFE as a cosolvent, and it was found that less reactive dipolarophiles such as 5b–g are also involved in the photocycloaddition process. In this way, a wide variety of novel cycloadducts could be obtained with moderate to good yields, such as thioester 7b, imide 7d, Weinreb amide 7e, amide 7f, and cyanide 7g (**Figure 1.25**).

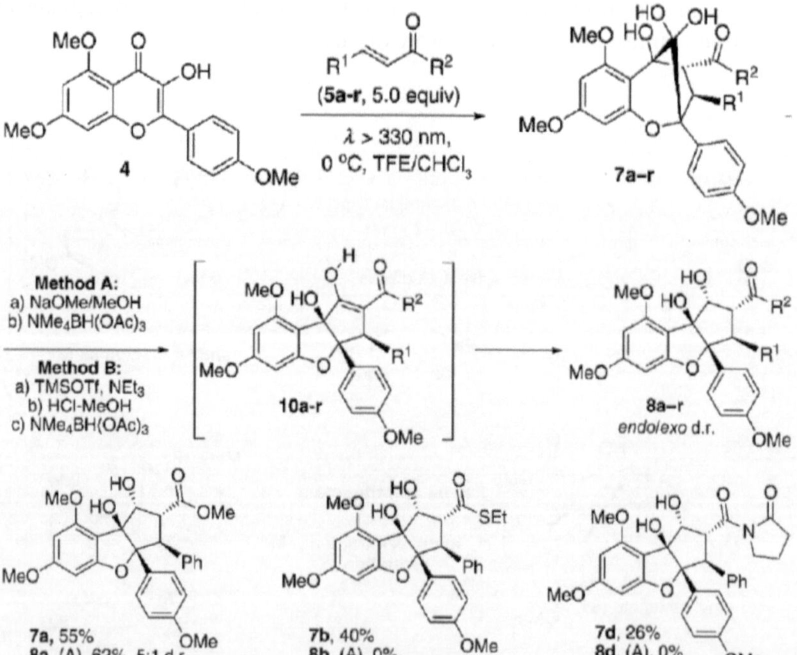

FIGURE 1.25   Synthesis of aglains 7a–r and rocaglates 8a–r. (Adapted with permission from ref. [124].) *(Continued)*

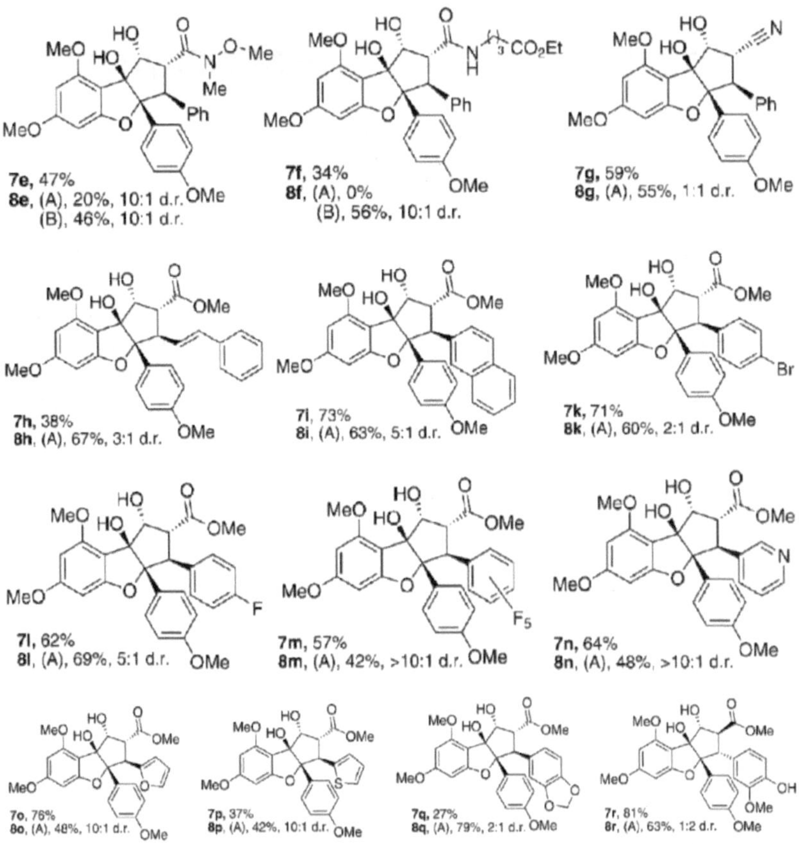

7e, 47%
8e, (A), 20%, 10:1 d.r.
(B), 46%, 10:1 d.r.

7f, 34%
8f, (A), 0%
(B), 56%, 10:1 d.r.

7g, 59%
8g, (A), 55%, 1:1 d.r.

7h, 38%
8h, (A), 67%, 3:1 d.r.

7i, 73%
8i, (A), 63%, 5:1 d.r.

7k, 71%
8k, (A), 60%, 2:1 d.r.

7l, 62%
8l, (A), 69%, 5:1 d.r.

7m, 57%
8m, (A), 42%, >10:1 d.r.

7n, 64%
8n, (A), 48%, >10:1 d.r.

7o, 76%
8o, (A), 48%, 10:1 d.r.

7p, 37%
8p, (A), 42%, 10:1 d.r.

7q, 27%
8q, (A), 79%, 2:1 d.r.

7r, 81%
8r, (A), 63%, 1:2 d.r.

FIGURE 1.25   *(Continued)*

Due to the availability of various rocaglate derivatives as racemic compounds, the potency of these rocaglate derivatives as eukaryotic translation inhibitors was compared with that of enantiopure silvestrol 2. Six out of 25 compounds tested were found to show greater than 50% inhibition of translation at 10 µm when evaluated for their potency in vitro; they were all endo cycloaddition diastereomers. In the titration of the six compounds, 8e and 8f were found to be the most potent inhibitors with IC50 values between 300 and 400 nM. IC_{50} for Silvestrol 2 was approximately hundred nanomolars, according to the same experiment. In order to determine the potency of the inhibitors, these analogues were tested in vivo for their ability to inhibit protein synthesis. Likewise, hydroxymate 8e inhibited 85% of protein synthesis over an hour, similar to silvestrol 2 that inhibited 85% of protein synthesis.

1.12 SYNTHESIS OF HAMIGERANS

Natural hamigerans 1–4 and their epimers were synthesized through the intramolecular Diels–Alder reaction of photochemically generated hydroxy-o-quinodimethanes. A suitable process has been developed for the construction of the benzannulated carbocyclic frameworks of hamigeran-class compounds (**Figure 1.26**), as immediate synthetic targets[125].

To synthesize the hamigerans, a second benzaldehyde was used, which had an oxygen functional at C6, in hopes that the additional handle would activate the necessary epimerization at C5. When 16 was photo-irradiated under standard conditions, a tricyclic hydroxy ester 17 was formed with a 92% yield and as a mixture of C10 epimers.

After heating 17 in methanolic HCl, both dehydration and MOM cleavage took place at 60°C, resulting in a 91% yield of the hydroxy unsaturated ester 18. By stereoselectively dihydroxylating the olefin 18, triol 19 was

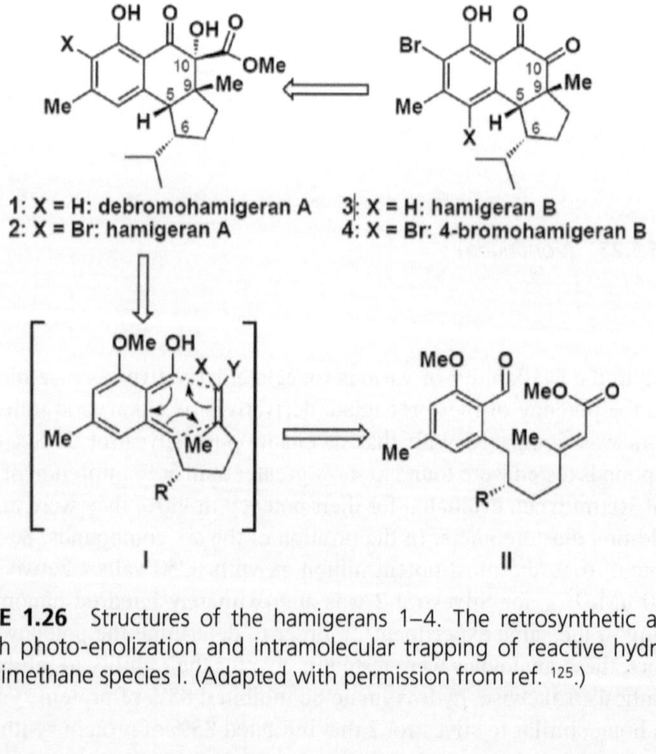

1: X = H: debromohamigeran A
2: X = Br: hamigeran A

3: X = H: hamigeran B
4: X = Br: 4-bromohamigeran B

I

II

FIGURE 1.26 Structures of the hamigerans 1–4. The retrosynthetic analysis through photo-enolization and intramolecular trapping of reactive hydroxy-o-quinodimethane species I. (Adapted with permission from ref. [125].)

obtained in a 91% yield by a known method. By selectively protecting the vicinal hydroxy groups in 19 with 2-methoxypropene and catalytic amounts of PPTS followed by Dess–Martin oxidation of the remaining hydroxy group, the corresponding ketone acetonide 21 was obtained.

Base-induced isomerization at C5 was extremely facile due to its neighboring carbonyl group, which led to the expected cis 5,6 junctions in product 21 after only 10 minutes of exposure to DBU at 0°C. Also, the carbonyl group also serves as an electrophile well, allowing the cerium-mediated introduction of the required isopropyl group (21–22), marking the arrival at intermediate 22. TFA/Et3SiH was used to reductively remove the tertiary alcohol group from 22 and resulted in the polycyclic ether 23. Since similar side reactions were observed under various other conditions, it was decided to produce olefin 24, which was readily obtained from alcohol 22 using SOCl$_2$/py and provided in an inseparable mixture with conjugated (24′) and exocyclic (24″) double bond isomers.

Nonetheless, obtaining the correct stereochemistry of hamigerans at C6 as expected from an exo face attack was a challenge. Several reduction experiments have shown that the suspected incorrect C6 stereoisomer 25 has been contaminated with unreactive tetrasubstituted olefinic isomers 24′ and 24″. Consequently, a change of course was required. Following that, it was decided to complete the syntheses of the new series of hamigeran analogues that could be produced from 24 as in the previous case. With HCl (3n) and THF (1:1), the acetonide group of the reduction product 25 was successfully hydrolyzed at 80°C during acid-induced hydrolysis, resulting in diol 26 (mixture of C11 epimers with approximately a 1.3:1 ratio), which could then be purified chromatographically and characterized spectroscopically.

The mixture of benzylic alcohols 26 was oxidized with SO$_3$.py/DMSO to yield an 82% yield of hydroxy ketoester 27, to obtain the single stereoisomer of the hydroxy ketoester 27. The demethylation of 27 led to the formation of phenol 28, which was then investigated by proton NMR spectroscopy (NOE) in order to confirm that the isopropyl group of 28 has the incorrect stereochemistry (C6), and that such a compound should be called 6-epidebromohamigeran A. By exposing the compound to a stoichiometric amount of NBS in the presence of iPr$_2$NH (95% yield), a smooth conversion of this compound into 6-epi-hamigeran A (29) was achieved.

A new branching sequence with readily available intermediates was considered as an alternative option in light of this new failure to establish the correct stereochemistry for C6. As shown in **Figure 1.27**, the first step in finding such a path had to be taken by starting with the olefin 24. The importance of this sequence was highlighted when olefin 24 was hydroborated

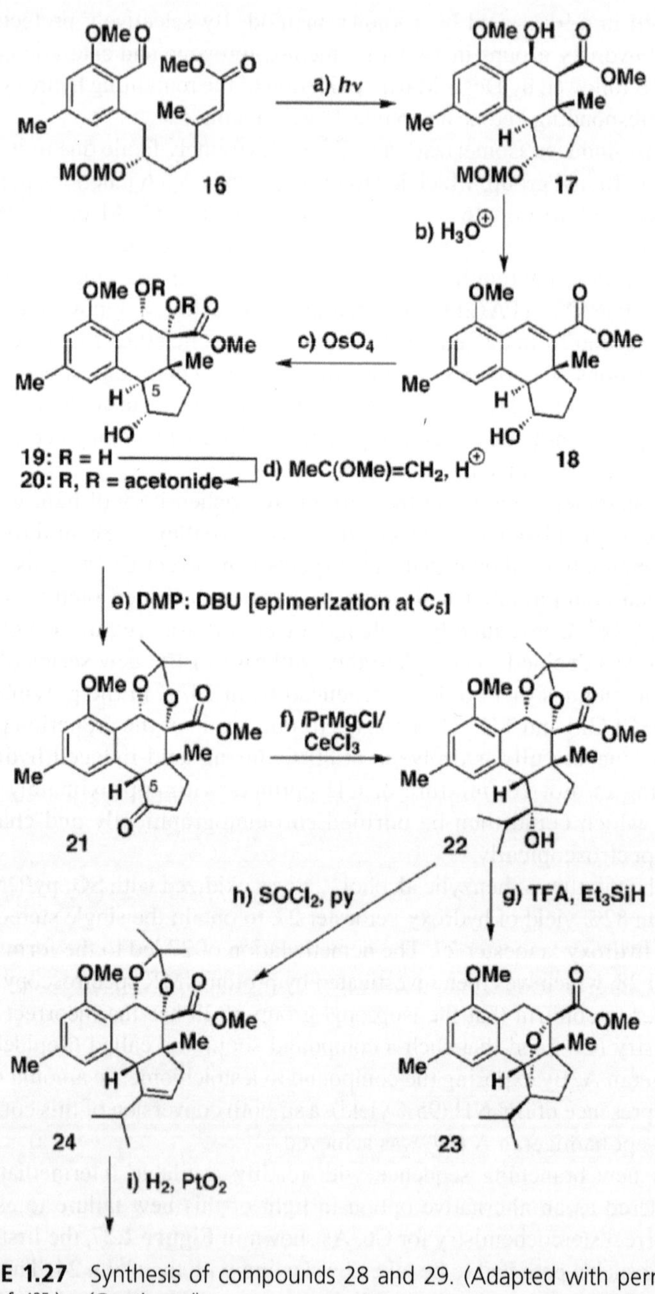

FIGURE 1.27   Synthesis of compounds 28 and 29. (Adapted with permission from ref. [125].) *(Continued)*

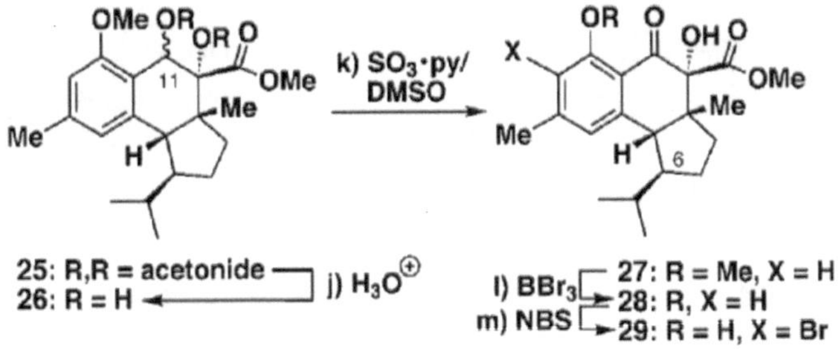

25: R,R = acetonide ⎤ j) H$_3$O$^{\oplus}$
26: R = H ◄──────────┘

27: R = Me, X = H
28: R, X = H
29: R = H, X = Br

l) BBr$_3$
m) NBS

FIGURE 1.27   *(Continued)*

with BH$_3$ mE$_2$S under sonication conditions to produce, following an oxidative process, the desired 6R,7R alcohol isomer 30 along with its a-stereoisomer. A chromatographic separation of isomer 30 resulted in the conversion of the isomer into phenylthionocarbonate (31) and then by heating with nBu$_3$SnH/AIBN, into deoxygenated product 32.

The creation of intermediate 32 allowed all four natural products to be synthesized. To obtain diol 33 in an 88% yield, the acetonide protecting group from 32 was removed by heating it at 80°C with HCl (1n) in THF (1:1). However, PDC was able to successfully convert 33 into 34, even though the standard SO$_3$ py/DMSO protocol yielded low conversion yields. By using the same BBr3-induced cleavage protocol used in the previous epi series, 34 was converted into debromohamigeran A (1) (95% yield). The second product, hamigeran A (2), was obtained through NBS-mediated bromination (95% yield) of debromohamigeran A (1).

A cascade reaction was developed a cascade reaction to convert hamigeran A (2) into hamigeran B (3) under aerobic conditions by introducing Ba(OH)$_2$ into MeOH/H$_2$O (2:1) followed by (1) saponification, (2) decarboxylation, and (3) auto-oxidation, yielding an overall yield of 82% (**Figure 1.28**). By facilitating the introduction of the second bromine atom using NBS, 4-bromohamigeran B (4) was synthesized from hamigeran B (3) (95% yield). By using dibromination (NBS, DMF) followed by treatment with Ba(OH)$_2$, 4-Bromohamigeran B (4) can also be obtained from 1 (65% overall yield).

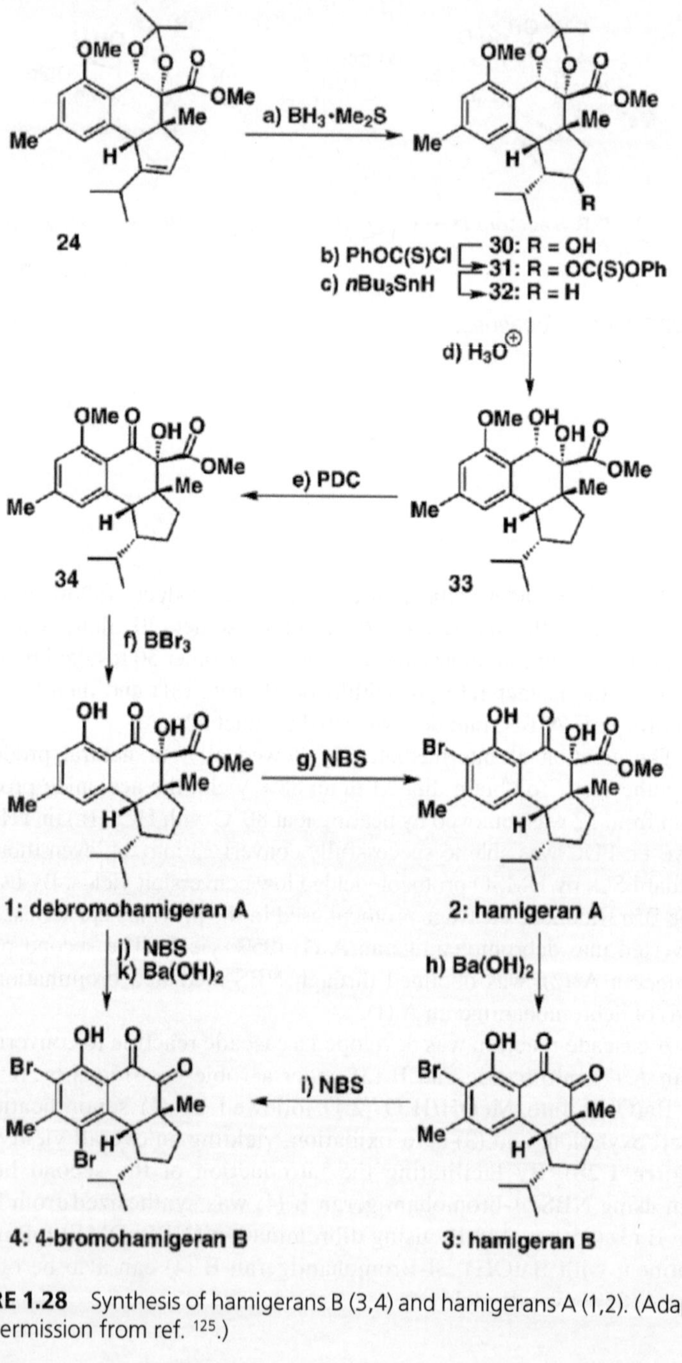

FIGURE 1.28 Synthesis of hamigerans B (3,4) and hamigerans A (1,2). (Adapted with permission from ref. [125].)

1.13 NATURALLY OCCURING XANTHANOLIDES AND THEIR SYNTHESIS

Scientists have already identified over a hundred members of a sesquiterpenoids family called xanthanolides[126]. Most xanthanolides are composed of a five/seven bicyclic system with a butyrolactone trans or cis attached to a seven-membered carbocyclic chain, as shown by xanthatin (1)[127,128] and 8-epixanthatin (2; Figure 1.29 a)[129,130]. A few structurally more complex congeners,

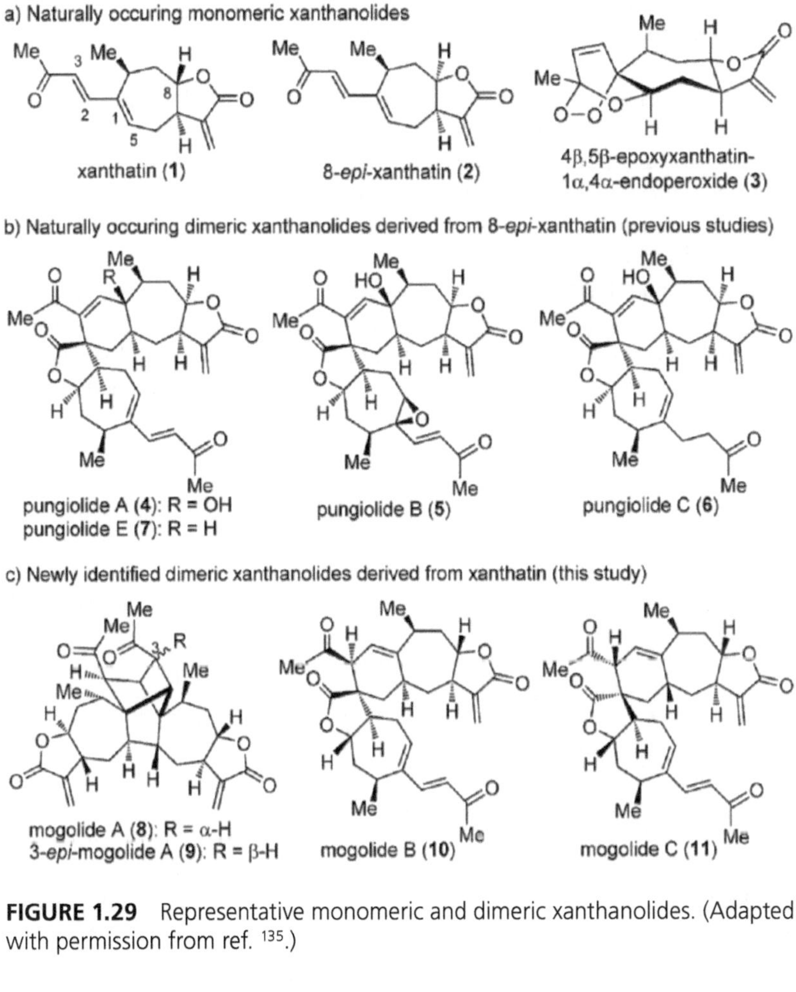

a) Naturally occuring monomeric xanthanolides

xanthatin (1)   8-epi-xanthatin (2)   4β,5β-epoxyxanthatin-1α,4α-endoperoxide (3)

b) Naturally occuring dimeric xanthanolides derived from 8-epi-xanthatin (previous studies)

pungiolide A (4): R = OH
pungiolide E (7): R = H   pungiolide B (5)   pungiolide C (6)

c) Newly identified dimeric xanthanolides derived from xanthatin (this study)

mogolide A (8): R = α-H
3-epi-mogolide A (9): R = β-H   mogolide B (10)   mogolide C (11)

FIGURE 1.29 Representative monomeric and dimeric xanthanolides. (Adapted with permission from ref. [135].)

like 4b,5b-epoxyxanthatin-1a,4a-endoperoxide (3; Figure 1.29 a) and pungiolides A–C (4–6)[131] and E (7; Figure 1.29 b)[132], have also been found in nature. It is not surprising that xanthanolides attract extensive attention from the synthetic community due to their intriguing chemical structures and diverse biological profiles[133,134].

Xanthanolide dimers were discovered through biomimetic synthesis[135]. Biomimetic synthesis of 4b,5b-epoxyxanthatin-1a,4a-endoperoxide, a novel monomeric compound, has been achieved using xanthatin as a starting material. Through three different dimerizations of xanthatin, up to four unprecedented xanthanolide dimers could also be synthesized. Head-to-head dimerizations or head-to-tail dimerizations were involved in these reactions. These dimeric compounds were initially identified as artifacts in the laboratory, but two of them, mogolides A and B, were later determined to be natural products found in Xanthium mogolium Kitag.

A dyotropic rearrangement of 3,4-cis-lactone was used to achieve the collective synthesis of monomeric xanthanolides. As a result of this strategy, even more challenging targets, such as 3–7, can be synthesized. As xanthanolides 3–7 have polycyclic skeletons that differ from those of other xanthanolides, their synthetic targets are challenging. For 3–7, there are also interesting biosynthesis origins. A possible biosynthetic pathway for 3-7 can be seen in **Figure 1.30**.

It was found that DBU prevented a significant amount of dimerization in the reaction. Consequently, the desired product 13a and its epimer 13b were obtained in a combined yield of 25% (**Figure 1.31**). Furthermore, by increasing the number of equivalents of DBU used, the yield of the reaction was

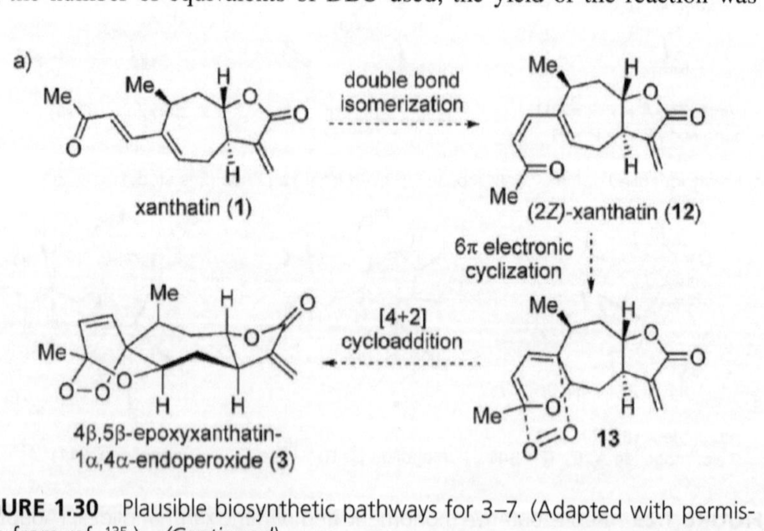

FIGURE 1.30 Plausible biosynthetic pathways for 3–7. (Adapted with permission from ref. [135].) *(Continued)*

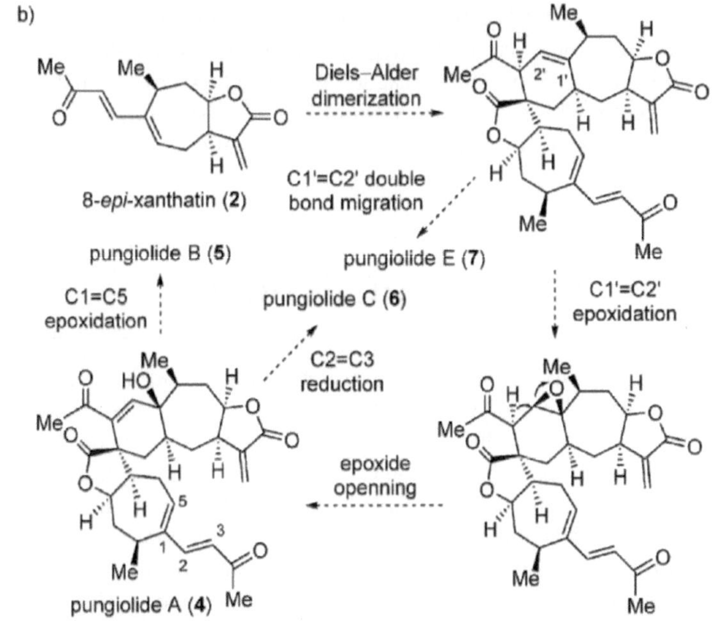

b)

FIGURE 1.30 *(Continued)*

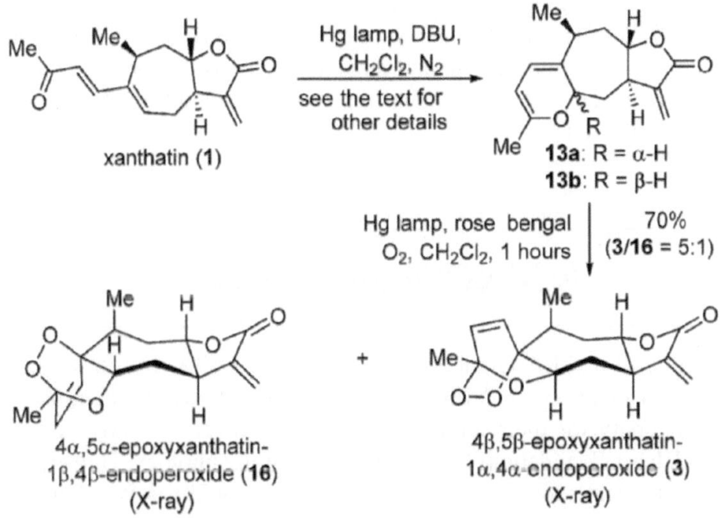

FIGURE 1.31 Biomimetic synthesis of 3. (Adapted with permission from ref. [135].)

improved by 60%, as well as the diastereoselectivity. Singletoxygen-promoted Diels–Alder reactions were performed on 13a and 13b under standard reaction conditions. In a 5:1 ratio, the reaction yielded a mixture of 3 and its diastereoisomer, 16, with a combined yield of 70%. The structures of 3 and 16 have been unambiguously confirmed by X-ray crystallography.

1.14 CONCLUSIONS

Recently, photocatalytic technology has shown great promise as a low-cost, environment-friendly, and sustainable technology. Using photochemistry to synthesize natural products has made great progress over the last few years. It is the purpose of this chapter to highlight some of those accomplishments. Photochemically mediated reactions have proven to be one of the most useful techniques for constructing highly complex molecular structures that would otherwise be difficult to construct using conventional, thermal processes due to their high strain and complexity during the past few decades. Potential practitioners who wish to use these procedures may find themselves hampered by the need to use specialized equipment in order to carry out the necessary photochemical reactions. While this may hinder the progress of photochemical transformations, there has been continuous progress in synthesizing natural products and biologically active compounds. The development and application of photochemistry in organic synthesis will continue for many years to come as further mechanistic studies clarify the general chemical reactivity of electronically generated excited intermediates.

ACKNOWLEDGMENTS

AD is grateful to CEA-Grenoble, Joseph Fourier University, University of Göttingen, University of California, Los Angeles, Prince Mohammad Bin Fahd University for their support. BKB is grateful to US NIH, US NCI, Texas Kleberg Foundation, Stevens Institute of Technology, University of Texas M. D. Anderson Cancer Center, University of Texas-Pan American, Community Health Systems of Texas, Prince Mohammad Bin Fahd University for their support.

REFERENCES

1. Trost BM. The atom economy – a search for synthetic efficiency. *Science.* 1991;254(5037):1471–1477. doi:10.1126/science.1962206
2. Keasling JD, Mendoza A, Baran PS. A constructive debate. *Nature.* 2012;492(7428):188–189. doi:10.1038/492188a
3. Mulzer J. Trying to rationalize total synthesis. *Nat Prod Rep.* 2014;31(4): 595–603. doi:10.1039/C3NP70105K
4. Paterson I, Anderson EA. Chemistry. The renaissance of natural products as drug candidates. *Science.* 2005;310(5747):451–453. doi:10.1126/science.1116364
5. Nicolaou KC, Hale CRH, Nilewski C, Ioannidou HA. Constructing molecular complexity and diversity: total synthesis of natural products of biological and medicinal importance. *Chem Soc Rev.* 2012;41(15):5185–5238. doi:10.1039/C2CS35116A
6. Newman DJ, Cragg GM, Snader KM. The influence of natural products upon drug discovery. *Nat Prod Rep.* 2000;17(3):215–234. doi:10.1039/A902202C
7. Newhouse T, Baran PS, Hoffmann RW. The economies of synthesis. *Chem Soc Rev.* 2009;38(11):3010–3021. doi:10.1039/B821200G
8. Grondal C, Jeanty M, Enders D. Organocatalytic cascade reactions as a new tool in total synthesis. *Nat Chem.* 2010;2(3):167–178. doi:10.1038/nchem.539
9. Nicolaou KC, Edmonds DJ, Bulger PG. Cascade reactions in total synthesis. *Angew Chem Int Ed.* 2006;45(43):7134–7186. doi:10.1002/anie.200601872
10. Davies HML, Sorensen EJ. Rapid complexity generation in natural product total synthesis. *Chem Soc Rev.* 2009;38(11):2981–2982. doi:10.1039/B918568M
11. Bach T, Hehn JP. Photochemical reactions as key steps in natural product synthesis. *Angew Chem Int Ed.* 2011;50(5):1000–1045.
12. Sheldon RA. Catalysis: the key to waste minimization. *J Chem Technol Biotechnol.* 1997;68(4):381–388. doi: https://doi.org/10.1002/(SICI)1097-4660 (199704)68:4<381::AID-JCTB620>3.0.CO;2-3
13. Lipshutz BH, Ghorai S. Transitioning organic synthesis from organic solvents to water. What's your E factor? *Green Chem.* 2014;16(8):3660–3679. doi:10.1039/C4GC00503A
14. Olivucci M, Santoro F. Chemical selectivity through control of excited-state dynamics. *Angew Chem Int Ed.* 2008;47(34):6322–6325. doi:10.1002/anie.200800898
15. Schapiro I, Melaccio F, Laricheva EN, Olivucci M. Using the computer to understand the chemistry of conical intersections. *Photochem Photobiol Sci.* 2011;10(6):867–886. doi:10.1039/C0PP00290A
16. Hoffmann N. Photochemical reactions of aromatic compounds and the concept of the photon as a traceless reagent. *Photochem Photobiol Sci.* 2012;11(11): 1613–1641. doi:10.1039/C2PP25074H
17. Turro NJ. Fun with photons, reactive intermediates, and friends. Skating on the edge of the paradigms of physical organic chemistry, organic supramolecular photochemistry, and spin chemistry. *J Org Chem.* 2011;76(24):9863–9890. doi:10.1021/jo201786a

18. De Keukeleire D, He SL. Photochemical strategies for the construction of polycyclic molecules. *Chem Rev.* 1993;93(1):359–380. doi:10.1021/cr00017a017
19. Stockdale TP, Williams CM. Pharmaceuticals that contain polycyclic hydrocarbon scaffolds. *Chem Soc Rev.* 2015;44(21):7737–7763. doi:10.1039/C4CS00477A
20. Das A. Recent developments in semipolar InGaN laser diodes. *Semiconductors.* 2021;55(2):272–282. doi:10.1134/S106378262102010X
21. Das A. A systematic exploration of InGaN/GaN quantum well-based light emitting diodes on semipolar orientations. *Opt Spectrosc.* 2022;130(3):137–149. doi:10.1134/S0030400X2203002X
22. Yadav RN, Hossain F, Das A, Srivastava AK, Banik BK. Organocatalysis: a recent development on stereoselective synthesis of o-glycosides. *Catal Rev.* 2022;66(1):1–118. doi:10.1080/01614940.2022.2041303
23. Das A, Yadav RN, Banik BK A novel Baker's yeast-mediated microwave-induced reduction of racemic 3-keto-2-azetidinones: facile entry to optically active hydroxy β-lactam derivatives. *Curr Organocatalysis.* 2022;9(2):195–198.
24. Das A, Banik BK Versatile synthesis of organic compounds derived from ascorbic acid. *Curr Organocatalysis.* 2022;9(1):14–33.
25. Das A, Banik BK. Dipole Moment Studies on Beta Lactams. In: Banik BK, ed. *Green Approaches in Medicinal Chemistry for Sustainable Drug Design.* Elsevier; 2024:523–542.
26. Das A, Banik BK. Dipole moment in medicinal research: green and sustainable approach. In: Banik BK, ed. *Green Approaches in Medicinal Chemistry for Sustainable Drug Design.* Elsevier; 2024:561–602.
27. Das A, Das A, Banik BK. Influence of dipole moments on the medicinal activities of diverse organic compounds. *J Indian Chem Soc.* 2021;98(2):100005. doi:10.1016/j.jics.2021.100005
28. Das A, Banik BK. β-lactams: geometry, dipole moment and anticancer activity. *J Indian Chem Soc.* 2020;97(11b(Nov 2020):2461–2467. doi:10.5281/zenodo. 5656689
29. Das A, Alqashqari AA, Banik BK. Quantum mechanical calculations of dipole moment of diverse imines. *J Indian Chem Soc.* 2021;97(9b):1563–1566.
30. Das A, Banik BK. Dipole moment studies on α-hydroxy-β-lactam derivatives. *J Indian Chem Soc.* 2021;97(9b):1567–1571.
31. Das A, Banik BK. Dipole moment and anticancer activity of beta lactams. *Indian J Pharm Sci.* 2021;83(5):1071–1074. doi:10.36468/pharmaceutical-sciences.862
32. Das A, Banik BK. Computational studies of physicochemical parameters on optically active anticancer β-lactams. *Heterocycl Lett.* 2023;13(1).
33. Das A, Yadav R, Banik BK. Dipole moment studies on anticancer polyaromatic compounds. *Heterocycl Lett.* 2024;14(2):287–291.
34. Das A, Banik BK. Studies on dipole moment of penicillin isomers and related antibiotics. *J Indian Chem Soc.* 2020;97(6):911–915.
35. Das A, Bose AK, Banik BK. Stereoselective synthesis of β-lactams under diverse conditions: unprecedented observations. *J Indian Chem Soc.* 2020; 97(6):917–925.
36. Das A, Banik BK. 26 - Dipole Moment in Medicinal Research: Green and Sustainable Approach. In: Banik BK, ed. *Green Approaches in Medicinal Chemistry for Sustainable Drug Design.* Advances in Green and Sustainable Chemistry. Elsevier; 2020:921–964. doi:10.1016/B978-0-12-817592-7.00021-6

37. Das A, Banik BK. *Microwaves in Chemistry Applications: Fundamentals, Methods and Future Trends.* Elsevier Science; 2021.

38. Das A, Banik BK. Chapter 1 - Foundational Principles of Microwave Chemistry. In: Das A, Banik B, eds. *Microwaves in Chemistry Applications.* Advances in Green and Sustainable Chemistry. Elsevier; 2021:3–26. doi:10.1016/B978-0-12-822895-1.00005-9

39. Das A, Banik BK. Chapter 2 - Microwave Equipment for Chemistry. In: Das A, Banik B, eds. *Microwaves in Chemistry Applications.* Advances in Green and Sustainable Chemistry. Elsevier; 2021:27–59. doi:10.1016/B978-0-12-822895-1.00002-3

40. Das A, Banik BK. Chapter 3 - Modeling and Interpreting Microwave Effects. In: Das A, Banik B, eds. *Microwaves in Chemistry Applications.* Advances in Green and Sustainable Chemistry. Elsevier; 2021:61–104. doi:10.1016/B978-0-12-822895-1.00007-2

41. Das A, Banik BK. Chapter 4 - Microwave-Assisted Synthesis of Oxygen- and Sulfur-Containing Organic Compounds. In: Das A, Banik B, eds. *Microwaves in Chemistry Applications.* Advances in Green and Sustainable Chemistry. Elsevier; 2021:107–142. doi:10.1016/B978-0-12-822895-1.00010-2

42. Das A, Banik BK. Chapter 5 - Microwave-Assisted Synthesis of N-Heterocycles. In: Das A, Banik B, eds. *Microwaves in Chemistry Applications.* Advances in Green and Sustainable Chemistry. Elsevier; 2021:143–198. doi:10.1016/B978-0-12-822895-1.00006-0

43. Das A, Banik BK. Chapter 6 - Microwave-Assisted Oxidation and Reduction Reactions. In: Das A, Banik B, eds. *Microwaves in Chemistry Applications.* Advances in Green and Sustainable Chemistry. Elsevier; 2021:199–244. doi:10.1016/B978-0-12-822895-1.00001-1

44. Das A, Banik BK. Chapter 7 - Microwave-Assisted Enzymatic Reactions. In: Das A, Banik B, eds. *Microwaves in Chemistry Applications.* Advances in Green and Sustainable Chemistry. Elsevier; 2021:245–281. doi:10.1016/B978-0-12-822895-1.00009-6

45. Das A, Banik BK. Chapter 8 - Microwave-Assisted Sterilization. In: Das A, Banik B, eds. *Microwaves in Chemistry Applications.* Advances in Green and Sustainable Chemistry. Elsevier; 2021:285–328. doi:10.1016/B978-0-12-822895-1.00011-4

46. Das A, Banik BK. Chapter 9 - Microwave-Assisted CVD Processes for Diamond Synthesis. In: Das A, Banik B, eds. *Microwaves in Chemistry Applications.* Advances in Green and Sustainable Chemistry. Elsevier; 2021:329–374. doi:10.1016/B978-0-12-822895-1.00004-7

47. Das A, Banik BK. Chapter 10 - Future Trends in Microwave Chemistry and Biology. In: Das A, Banik B, eds. *Microwaves in Chemistry Applications.* Advances in Green and Sustainable Chemistry. Elsevier; 2021:375–384. doi:10.1016/B978-0-12-822895-1.00003-5

48. Das A, Yadav RN, Banik BK. Microwave-induced conversion of electromagnetic energy into heat energy in different solvents: Synthesis of beta lactams. *Chem J Mold.* 2022;17(1):62–66. doi:10.19261/cjm.2021.864

49. Das A, Banik BK. Microwave-induced biocatalytic reactions toward medicinally important compounds. *Phys Sci Rev.* 2022;7(4-5):507–538. doi:10.1515/psr-2021-0064

50. Das A, Banik BK 3 Microwave-Induced Biocatalytic Reactions Toward Medicinally Important Compounds. In: *3 Microwave-Induced Biocatalytic Reactions Toward Medicinally Important Compounds.* De Gruyter; 2022:57–88. doi:10.1515/9783110732542-003

51. Das A, Yadav R, Banik B. Microwave-induced surface-mediated highly efficient regioselective nitration of aromatic compounds: effects of penetration depth. *Asian J Chem.* 2021;33:2203–2206. doi:10.14233/ajchem.2021.23131

52. Das A, Banik BK. Microwave-induced catalytic transfer hydrogenation in different solvents toward optically active hydroxy beta lactams: effects of penetration depth. *Asian J Org Med Chem.* 2023;8(1):7–10.

53. Das A, Banik BK. Microwave in research-more miracles. Heterocycl Lett. 2024;14(2):449–456.

54. Das A, Banik BK. Expeditious synthesis of oxygen and sulfur heterocycles by microwave. Heterocycl Lett. 2024;14(2):457–467.

55. Das A, Yadav R, Banik BK. Microwave-induced ferrier rearrangement of hyroxy beta-lactams with glycals. *Appl Chem Eng.* 2024;7(2):1870–1870.

56. Aramaki Y, Chiba K, Tada M. Spiro-lactones, hyperolactone A-D from *Hypericum chinense. Phytochemistry.* 1995;38(6):1419–1421. doi:10.1016/0031-9422(94)00862-N

57. Nicolaou KC, Sarlah D, Shaw DM. Total synthesis and revised structure of biyouyanagin A. *Angew Chem Int Ed.* 2007;46(25):4708–4711. doi:10.1002/anie.200701552

58. Tanaka N, Okasaka M, Ishimaru Y, Takaishi Y, Sato M, Okamoto M, et al. biyouyanagin A, an anti-HIV agent from *Hypericum chinense* L. var. salicifolium. *Org Lett.* 2005;7(14):2997–2999. doi:10.1021/ol050960w

59. Mangion IK, MacMillan DWC. Total synthesis of brasoside and littoralisone. *J Am Chem Soc.* 2005;127(11):3696–3697. doi:10.1021/ja050064f

60. Das A. LED light sources in organic synthesis: an entry to a novel approach. *Lett Org Chem.* 2022;19(4):283–292.

61. Das A, Banik BK. Sustainable reactions in the synthesis of heterocycles. *Curr Organocatalysis.* 2022;9(1):3–3. doi:10.2174/2213337209012203281645 23

62. Yadav RN, Shaikh AL, Das A, Ray D, Banik BK. Asymmetric synthesis of 3-pyrrole substituted β-lactams through p-toluene sulphonic acid-catalyzed reaction of azetidine-2,3-diones with hydroxyprolines. *Curr Organocatalysis.* 2022;9(4):337–345.

63. Das A, Yadav RN, Banik BK. Conceptual design and cost-efficient environmentally benign synthesis of beta-lactams. *Phys Sci Rev.* 2022;8(11):4053–4084. Published online May 4, 2022. doi:10.1515/psr-2021-0088

64. Das A, Yadav R, Banik BK. 10 Conceptual Design and Cost-Efficient Environmentally Benign Synthesis of Betalactams. In: *10 Conceptual Design and Cost-Efficient Environmentally Benign Synthesis of Betalactams.* De Gruyter; 2022:357–388. doi:10.1515/9783110797428-010

65. Anderson JR, Edwards RL, Poyser JP, Whalley AJS. Metabolites of the higher fungi. Part 23. The punctaporonins. Novel bi-, tri-, and tetra-cyclic sesquiterpenes related to caryophyllene, from the fungus Poronia punctata (Linnaeus: Fries) Fries. *J Chem Soc, Perkin Trans 1.* 1988;(4):823–831. doi:10.1039/P19880000823

66. Anderson JR, Edwards RL, Freer AA, Mabelis RP, Poyser JP, Spencer H, et al. Punctatins B and C (antibiotics M95154 and M95155): further sesquiterpene

alcohols from the fungus *Poronia punctata*. *J Chem Soc, Chem Commun.* 1984;(14):917–919. doi:10.1039/C39840000917

67. Fleck M, Bach T. Total synthesis of the tetracyclic sesquiterpene (±)-punctaporonin c. *Angew Chem Int Ed.* 2008;47(33):6189–6191. doi:10.1002/anie.200801534

68. Srikrishna A, Ramasastry SSV. Enantiospecific total synthesis of phytoalexins, (+)-solanascone, (+)-dehydrosolanascone, and (+)-anhydro-β-rotunol. *Tetrahedron Lett.* 2005;46(43):7373–7376. doi:10.1016/j.tetlet.2005.08.124

69. Takao K, Kai H, Yamada A, Fukushima Y, Komatsu D, Ogura A, et al. Total syntheses of (+)-aquatolide and related humulanolides. *Angew Chem Int Ed.* 2019;58(29):9851–9855. doi:10.1002/anie.201904404

70. Kleinnijenhuis RA, Timmer BJJ, Lutteke G, Smits JMM, de Gelder R, van Maarseveen JH, et al. Formal synthesis of solanoeclepin A: enantioselective allene diboration and intramolecular [2+2] photocycloaddition for the construction of the tricyclic core. *Chemistry.* 2016;22(4):1266–1269. doi:10.1002/chem.201504894

71. Riener M, Nicewicz DA. Synthesis of cyclobutane lignans via an organic single electron oxidant–electron relay system. *Chem Sci.* 2013;4(6):2625–2629. doi:10.1039/C3SC50643F

72. Hurtley AE, Lu Z, Yoon TP. [2+2] Cycloaddition of 1,3-dienes by visible light photocatalysis. *Angew Chem Int Ed.* 2014;53(34):8991–8994. doi:10.1002/anie.201405359

73. Lu Z, Yoon TP. Visible light photocatalysis of [2+2] styrene cycloadditions by energy transfer. *Angew Chem Int Ed.* 2012;51(41):10329–10332. doi:10.1002/anie.201204835

74. Das A, Banik BK. Tellurium-based solar cells. *Phys Sci Rev.* 2022;8(12):4631–4658. Published online May 18, 2022. doi:10.1515/psr-2021-0110

75. Das A, Banik BK. 5 Tellurium-Based Solar Cells. In: *5 Tellurium-Based Solar Cells.* De Gruyter; 2022:107–134. doi:10.1515/9783110735840-005

76. Das A, Banik BK. Semiconductor characteristics of tellurium and its implementations. *Phys Sci Rev.* 2022;8(12):4659–4687. Published online May 18, 2022. doi:10.1515/psr-2021-0108

77. Das A, Banik BK. 3 Semiconductor Characteristics of Tellurium and Its Implementations. In: *3 Semiconductor Characteristics of Tellurium and Its Implementations.* De Gruyter; 2022:55–84. doi:10.1515/9783110735840-003

78. Das A, Das A, Banik BK. Tellurium-based chemical sensors. *Phys Sci Rev.* 2022;8(12):4461–4501. Published online May 17, 2022. doi:10.1515/psr-2021-0116

79. Das A, Das A, Banik BK. 9 Tellurium-Based Chemical Sensors. In: *9 Tellurium-Based Chemical Sensors.* De Gruyter; 2022:183–224. doi:10.1515/9783110735840-009

80. Das A, Ray D, Banik BK. Tellurium in carbohydrate synthesis. *Phys Sci Rev.* 2022;8(11):4157–4178. Published online May 7, 2022. doi:10.1515/psr-2021-0109

81. Das A, Ray D, Banik BK. 4 Tellurium in Carbohydrate Synthesis. In: *4 Tellurium in Carbohydrate Synthesis.* De Gruyter; 2022:85–106. doi:10.1515/9783110735840-004

82. Aldawood SAA, Das A, Banik BK. Tellurium-induced cyclization of olefinic compounds. *Phys Sci Rev.* 2022;8(12):4569–4609. Published online May 17, 2022. doi:10.1515/psr-2021-0119

83. Aldawood SAA, Das A, Banik BK. 11 Tellurium-Induced Cyclization of Olefinic Compounds. In: *11 Tellurium-Induced Cyclization of Olefinic Compounds.* De Gruyter; 2022:249–290. doi:10.1515/9783110735840-011

84. Ray D, Das A, Mazumdar S, Banik BK. Tellurium-induced functional group activation. *Phys Sci Rev.* 2023;8(12):4821–4838. Published online June 2, 2022. doi:10.1515/psr-2021-0221

85. Ray D, Das A, Mazumdar S, Banik BK. 12 Tellurium-Induced Functional Group Activation. In: *12 Tellurium-Induced Functional Group Activation.* De Gruyter; 2022:291–308. doi:10.1515/9783110735840-012

86. Furst L, Narayanam JMR, Stephenson CRJ. Total synthesis of (+)-gliocladin c enabled by visible-light photoredox catalysis. *Angew Chem Int Ed.* 2011; 50(41):9655–9659. doi:10.1002/anie.201103145

87. Hino T, Nakagawa M. Chapter 1 Chemistry and Reactions of Cyclic Tautomers of Tryptamines and Tryptophans. In: Brossi A, ed. *The Alkaloids: Chemistry and Pharmacology.* Vol 34. Academic Press; 1989:1–75. doi:10.1016/S0099-9598(08)60226-6

88. Usami Y, Yamaguchi J, Numata A. Gliocladins A—C and glioperazine: Cytotoxic dioxo- or trioxopiperazine metabolites from a *Gliocladium* sp. separated from a sea hare. *ChemInform.* 2004;35(36). doi:10.1002/chin.200436204

89. Dong JY, He HP, Shen YM, Zhang KQ. Nematicidal epipolysulfanyldioxo-piperazines from Gliocladium roseum. *J Nat Prod.* 2005;68(10):1510–1513. doi:10.1021/np0502241

90. Takahashi C, Numata A, Matsumura E, Minoura K, Eto H, Shingu T, et al. Leptosins I and J, cytotoxic substances produced by a *Leptosphaeria* sp. Physico-chemical properties and structures. *J Antibiot (Tokyo).* 1994;47(11):1242–1249. doi:10.7164/antibiotics.47.1242

91. Zheng CJ, Kim CJ, Bae KS, Kim YH, Kim WG. Bionectins A-C, epidithiodi-oxopiperazines with anti-MRSA activity, from *Bionectra byssicola* F120. *J Nat Prod.* 2006;69(12):1816–1819. doi:10.1021/np060348t

92. Das A, Banik BK. 15 - Versatile Thiosugars in Medicinal Chemistry. In: Banik BK, ed. *Green Approaches in Medicinal Chemistry for Sustainable Drug Design.* Advances in Green and Sustainable Chemistry. Elsevier; 2020: 549–574. doi:10.1016/B978-0-12-817592-7.00015-0

93. Das A, Banik BK. Versatile Thiosugars in Medicinal Chemistry. In: Banik BK, ed. *Green Approaches in Medicinal Chemistry for Sustainable Drug Design.* Elsevier; 2024:409–441.

94. Das A, Yadav RN, Banik BK. Ascorbic acid-mediated reactions in organic synthesis. *Curr Organocatalysis.* 2020;7(3):212–241.

95. Das A, Banik BK. Graphene Oxide and Modified Graphene Oxide-Mediated Synthesis of Medicinally Active Compounds. In: *Green Approaches in Medicinal Chemistry for Sustainable Drug Design.* Elsevier; 2024:13–44.

96. Das A, Banik BK. Synthesis of Natural Products by Photochemistry. In: Banik BK, ed. *Green Approaches in Medicinal Chemistry for Sustainable Drug Design.* Elsevier; 2024:259–283.

97. Das A, Banik BK. Green Synthesis of Biologically Active pyrroles and related substrates via C-H Functionalization. In: Banik BK, ed. *Green Approaches in Medicinal Chemistry for Sustainable Drug Design.* Elsevier; 2024:101–131.

98. Das A, Ashraf MW, Banik BK. Thione derivatives as medicinally important compounds. *ChemistrySelect.* 2021;6(34):9069–9100. doi:10.1002/slct.202102398

99. López CS, Pérez-Balado C, Rodríguez-Graña P, de Lera AR. Mechanistic insights into the stereocontrolled synthesis of hexahydropyrrolo[2,3-b]indoles by electrophilic activation of tryptophan derivatives. *Org Lett.* 2008;10(1): 77–80. doi:10.1021/ol702732j

100. Beatty JW, Stephenson CRJ. Synthesis of (−)-pseudotabersonine, (−)-pseudovincadifformine, and (+)-coronaridine enabled by photoredox catalysis in flow. *J Am Chem Soc.* 2014;136(29):10270–10273. doi:10.1021/ja506170g

101. Gerard B, Jones G, Porco JA. A biomimetic approach to the rocaglamides employing photogeneration of oxidopyryliums derived from 3-hydroxyflavones. *J Am Chem Soc.* 2004;126(42):13620–13621. doi:10.1021/ja0447980

102. Proksch P, Edrada R, Ebel R, Bohnenstengel FI, Nugroho BW. Chemistry and biological activity of rocaglamide derivatives and related compounds in *Aglaia* species (Meliaceae). *Current Organic Chemistry.* 5(9):923–938. doi:10.2174/1385272013375049

103. Salim AA, Pawlus AD, Chai HB, Farnsworth NR, Douglas Kinghorn A, Carcache-Blanco EJ. Ponapensin, a cyclopenta[bc]benzopyran with potent NF-κB inhibitory activity from *Aglaia ponapensis. Bioorg Med Chem Lett.* 2007;17(1):109–112. doi:10.1016/j.bmcl.2006.09.084

104. Bordeleau ME, Mori A, Oberer M, Lindqvist L, Chard LS, Higa T, et al. Functional characterization of IRESes by an inhibitor of the RNA helicase eIF4A. *Nat Chem Biol.* 2006;2(4):213–220. doi:10.1038/nchembio776

105. Trost BM, Greenspan PD, Yang BV, Saulnier MG. An unusual oxidative cyclization. A synthesis and absolute stereochemical assignment of (-)-rocaglamide. *J Am Chem Soc.* 1990;112(24):9022–9024. doi:10.1021/ja00180a081

106. Das A, Banik BK. Advances in heterocycles as DNA intercalating cancer drugs. *Phys Sci Rev.* Published online January 5, 2022;8 (9):2473–2521. doi:10.1515/psr-2021-0065

107. Das A, Banik BK. 4 Advances in heterocycles as DNA intercalating cancer drugs. In: *Heterocyclic Anticancer Agents.* De Gruyter; 2022:111–160. doi:10.1515/9783110735772-004

108. Banik BK, Das A. *Natural Products as Anticancer Agents.* Elsevier Science; 2023.

109. Banik BK, Das A. Anticancer Activity of Natural Compounds from Marine plants. In: Banik BK, Das A, eds. *Natural Products as Anticancer Agents.* Elsevier; 2024:237–284.

110. Banik BK, Das A. Anticancer Activity of Natural Compounds from Bacteria. In: Banik BK, Das A, eds. *Natural Products as Anticancer Agents.* Elsevier; 2024:287–328.

111. Banik BK, Das A. Anticancer Activity of Natural Compounds from Fungi. In: Banik BK, Das A, eds. *Natural Products as Anticancer Agents.* Elsevier; 2024:329–366.

112. Banik BK, Das A. Anticancer Drugs from Hormones and Vitamins. In: Banik BK, Das A, eds. *Natural Products as Anticancer Agents.* Elsevier; 2024:369–414.

113. Banik BK, Das A. Future Prospect in Anticancer Natural Products. In: Banik BK, Das A, eds. *Natural Products as Anticancer Agents.* Elsevier; 2024:415–426.

114. Das A, Banik BK. Anticancer Activity of Natural Compounds from Leaves of the Plants. In: Banik BK, Das A, eds. *Natural Products as Anticancer Agents.* Elsevier; 2024:3–48.
115. Das A, Banik BK. Anticancer Activity of Natural Compounds from Stems/ Barks of the Plants. In: Banik BK, Das A, eds. *Natural Products as Anticancer Agents.* Elsevier; 2024:49–86.
116. Das A, Banik BK. Anticancer Activity of Natural Compounds from Roots of the Plants. In: Banik BK, Das A, eds. *Natural Products as Anticancer Agents.* Elsevier; 2024:87–132.
117. Das A, Banik BK. Anticancer Activity of Natural Compounds from Fruits and Vegetables. In: Banik BK, Das A, eds. *Natural Products as Anticancer Agents.* Elsevier; 2024:133–178.
118. Das A, Banik BK. Anticancer Activity of Natural Compounds from Marine Animals. In: Banik BK, Das A, eds. *Natural Products as Anticancer Agents.* Elsevier; 2024:181–236.
119. Das A, Banik BK. Combatting the coronavirus utilizing natural cinnamon and its derived products. *Asian J Synth Nat Prod Chem.* 2023;1(1):11–15.
120. Das A. Quantitative structure-property relationships of taxol, taxotere and their epi-isomers. *J Indian Chem Soc.* 2020;97(11):9.
121. Shaikh AL, Das A, Banik BK. Indium-mediated reduction of aromatic nitro groups in β-lactams to oxazines. Heterocycl lett. 2024;14(2):267–272.
122. Gerard B, Cencic R, Pelletier J, Porco JA. Enantioselective synthesis of the complex rocaglate (-)-silvestrol. *Angew Chem Int Ed Engl.* 2007;46(41): 7831–7834. doi:10.1002/anie.200702707
123. Gerard B, Sangji S, O'Leary DJ, Porco JA. Enantioselective photocycloaddition mediated by chiral Brønsted acids: Asymmetric synthesis of the rocaglamides. *J Am Chem Soc.* 2006;128(24):7754–7755. doi:10.1021/ja062621j
124. Roche SP, Cencic R, Pelletier J, Porco JA. Biomimetic photocycloaddition of 3-hydroxyflavones: synthesis and evaluation of rocaglate derivatives as inhibitors of eukaryotic translation. *Angew Chem Int Ed Engl.* 2010;49(37): 6533–6538. doi:10.1002/anie.201003212
125. Nicolaou KC, Gray D, Tae J. Total synthesis of hamigerans: part 2. Implementation of the intramolecular Diels–Alder trapping of photochemically generated hydroxy-o-quinodimethanes; strategy and completion of the synthesis. *Angew Chem Int Ed.* 2001;40(19):3679–3683. doi: https://doi.org/10.1002/1521-3773(20011001)40:19<3679::AID-ANIE3679>3.0.CO;2-T
126. Vasas A, Hohmann J. Xanthane sesquiterpenoids: structure, synthesis and biological activity. *Nat Prod Rep.* 2011;28(4):824–842. doi:10.1039/C0NP00011F
127. Marco JA, Sanz-Cervera JF, Corral J, Carda M, Jakupovic J. Xanthanolides from *Xanthium:* absolute configuration of xanthanol, isoxanthanol and their C-4 epimers. *Phytochemistry.* 1993;34(6):1569–1576. doi:10.1016/S0031-9422(00)90847-1
128. Bohlmann F, Knoll KH, El-Emary NA. Neuartige sesquiterpenlactone aus *Pulicaria crispa. Phytochemistry.* 1979;18(7):1231–1233. doi:10.1016/0031-9422(79)80146-6
129. Nour AMM, Khalid SA, Kaiser M, Brun R, Abdallah WE, Schmidt TJ. The antiprotozoal activity of sixteen Asteraceae species native to Sudan and bioactivity-guided isolation of xanthanolides from *Xanthium brasilicum. Planta Med.* 2009;75(12):1363–1368. doi:10.1055/s-0029-1185676

130. Ahmed AA, Jakupovic J, Bohlmann F, Regaila HA, Ahmed AM. Sesquiterpene lactones from *Xanthium pungens*. *Phytochemistry*. 1990;29(7):2211–2215. doi:10.1016/0031-9422(90)83040-8

131. Ahmed AA, Mahmoud AA, El-Gamal AA. A xanthanolide diol and a dimeric xanthanolide from *Xanthium* species. *Planta Med*. 1999;65(5):470–472. doi:10.1055/s-2006-960817

132. Wang L, Wang J, Li F, Liu X, Chen B, Tang YX, et al. Cytotoxic sesquiterpene lactones from aerial parts of *Xanthium sibiricum*. *Planta Med*. 2013;79(8): 661–665. doi:10.1055/s-0032-1328482

133. Kummer DA, Brenneman JB, Martin SF. Application of a domino intramolecular enyne metathesis/cross metathesis reaction to the total synthesis of (+)-8-epi-xanthatin. *Org Lett*. 2005;7(21):4621–4623. doi:10.1021/ol051711a

134. Evans MA, Morken JP. Asymmetric synthesis of (–)-dihydroxanthatin by the stereoselective Oshima–Utimoto reaction. *Org Lett*. 2005;7(15):3371–3373. doi:10.1021/ol051276k

135. Shang H, Liu J, Bao R, Cao Y, Zhao K, Xiao C, et al. Biomimetic synthesis: discovery of xanthanolide dimers. *Angew Chem Int Ed*. 2014;53(52):14494–14498. doi:10.1002/anie.201406461

Solar Light-Mediated Synthesis of Biologically Active Molecules

2

Bimal Krishna Banik and Aparna Das

2.1 INTRODUCTION

Researchers have been working on developing solar-powered photocatalysis strategies over the past few years. This field has produced numerous outstanding studies that have demonstrated the infinite accessibility, environmental benignity, and mildness of visible spectrum. Additionally, visible light-promoted transformations are extremely selective and tolerant, as unplanned side reactions of compounds that are ultraviolet sensitive can be avoided. Using visible light as a light source, photoredox catalysts become photoexcited easily, which allows them to be quenched both oxidatively and reductively. Photoredox catalysis with visible light is an effective method of generating reactive radical intermediates through single electron transfer (SET). A number of artificial light sources have also been used to conduct photocatalysis, including lamps, lasers[1], and light-emitting diodes (LEDs)[2]. We have found this topic to be very useful since we are involved in conducting

DOI: 10.1201/9781003634249-2

multiple catalytic reactions[3–5]. Through cycloaddition reactions using thermal, photochemical, and microwave-induced irradiation methods, our group has prepared numerous beta-lactams[6–17]. A variety of organic structures are prepared by our group using microwave-induced energy[18–36].

The field of light-induced chemical reactions is receiving increasing attention from both industry and academia since it provides the energy required for chemical reactions to take place under mild conditions without creating any byproducts that are harmful to nature. In the past, ultraviolet (UV) light has been used to activate the reagents, but the processes have been difficult to operate because these reactions require expensive equipment, are difficult to control, and are not particularly selective. Visible light can be handled more easily than high-energy UV light. Furthermore, complex organic molecules are more resistant to photodecomposition when exposed to visible light, as opposed to UV light of higher energy. Organic chemistry has long and fascinating histories of using visible light photoredox catalysis (VLPC)[37–42]. Various chemical reactions have been conducted by our research group using microwave-induced energy[18–33,43].

Solar-activated photocatalysts (PCs) can also be used to induce environmentally friendly and reliable chemical synthesis[44]. The use of photocatalysis in organic synthesis allows for the initiation of conversions through many mechanistic approaches, including proton-coupled electron transfers, hydrogen atom transfers, and photocatalysis. Its simplicity and efficiency have made solar light promote photocatalytic redox a widely recognized method in organic synthesis. In recent decades, solar energy has received tremendous attention as an abundant, clean, affordable, and renewable means of remediating contaminated wastewater[45,46]. A few examples of solar light-induced synthesis of biologically significant compounds are presented in this chapter.

2.2 SOLAR LIGHT-INDUCED REACTIONS FOR THE SYNTHESIS OF CYCLIC COMPOUNDS

Cyclopropane is one of the highly strained three-membered carbocyclic rings that exhibit a high level of reactivity toward nucleophilic reagents. Natural and synthetic cyclopropanes bear a wide range of biological properties, including enzyme inhibition, insecticide, herbicide, antibiotic, antimicrobial, antitumor, antibacterial, and antiviral properties. The compounds containing

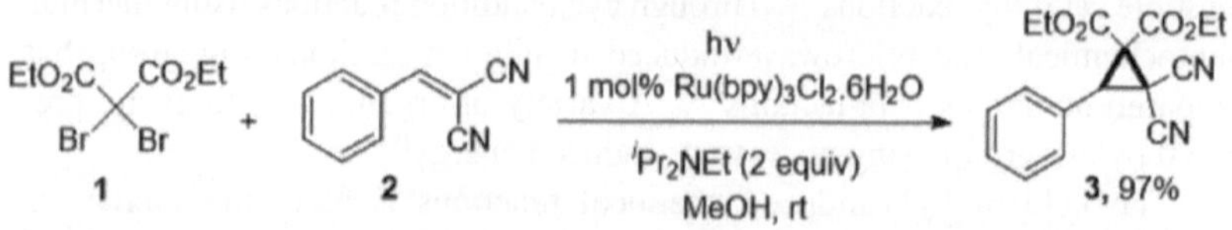

FIGURE 2.1 Cyclopropanation of dibromomalonates with alkenes. (Adapted with permission[54].)

cyclopropane are therefore of interest to chemical researchers in general, especially in organic and bioorganic chemistry. A number of medicinally active compounds have been studied by our research group. A number of novel sugars, natural products, and carbocyclic and heterocyclic organic compounds have been synthesized as a result of these efforts[5,47–53].

With photoredox catalysis, it has been demonstrated that dibromomalonates can be cyclopropanated with alkenes by a double electron transfer reaction induced by visible light[54]. A cyclopropanation protocol has been developed using 2-benzylidenemalononitrile **2** and diethyl 2,2-dibromomalonate **1** as substrates (**Figure 2.1**). It was a smooth reaction at the time of the initial attempt. The effectiveness of these reactions was then determined by performing control reactions. When no light is present during the reaction, either with or without a catalyst, the reaction occurs very slowly, resulting in very low yields of cyclopropane **3** after the starting material **1** has been converted. Under the influence of sunlight, efficiency and yield were significantly increased. In order to investigate this photocyclopropanation in more detail, a series of alkenes along with dibromomalonate derivatives have been investigated under optimal conditions.

Using visible light for double SET, carbanion generation was demonstrated to be effective for cyclopropanation. Since the conditions were mild and the synthetic method was simple, the cyclopropane derivatives synthesized using sunlight in open air and mild conditions are both environmentally benign and easy to handle.

Benzene rings are fused together to form the ortho-fused polycyclic arene picene. A great deal of therapeutic potential and bioactivity can be found in picene derivatives such as triterpenoids and pentacyclic triterpenes. Numerous pentacyclic triterpenes, including ursane, oleanolic acid, and oleane, have been reported to possess antitumor[55,56], antiviral, antidiabetic[57], and anti-inflammatory properties[58]. Moreover, the anti-inflammatory and anti-cancer properties of octadecahydropicene-2,3,14,15-tetranone obtained from Ziziphus nummularia root bark have been demonstrated[59,60]. The method and compounds reported in this investigation have similar functions to those reported in our research about anticancer molecules[61–76]. The development of

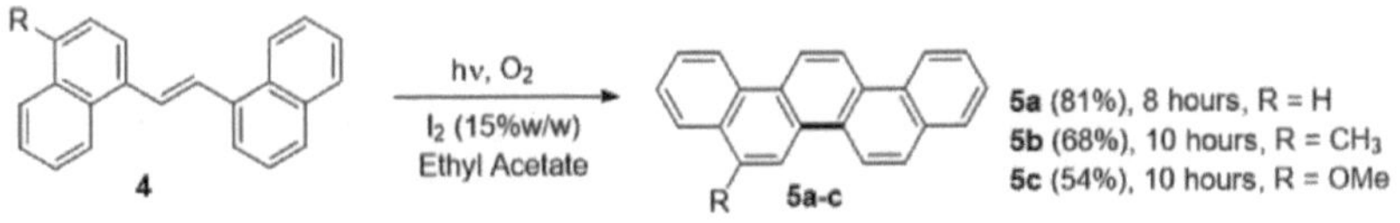

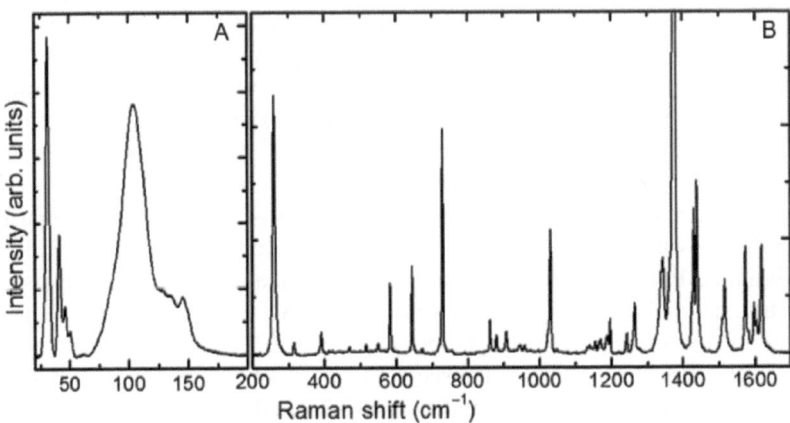

FIGURE 2.2 Synthesis of picenes. (Adapted with permission from[77].)

a synthetic protocol for substituted picene has proven to be one of the most challenging aspects of this study. Mallory photocyclization induced by solar light has been reported to produce substituted picenes[77].

In a straightforward Mallory reaction, dinaphthylethenes are photocyclized by solar light to produce picenes (**Figure 2.2**). As an alternative to benzene, various media were tested to determine whether they would be safe for the reaction. After the reaction mixture was exposed to sunlight, the product was easily recoverable through a simple filtration procedure. A Raman spectroscopy analysis (**Figure 2.3**) was performed on the polyene as well as X-ray diffraction to characterize its properties.

A titanium dioxide (TiO_2) surface modified with highly dispersed NiO particles exhibits greater absorption in the visible spectral region compared to an unmodified TiO_2 surface. There is also a reduced recombination of hole-electron pairs. By using these heterogeneous PCs, tertiary anilines

FIGURE 2.3 (A) Raman spectrum of solid picene in the low-frequency region characterized by intermolecular vibrations. (B) Raman spectrum of solid picene in the high-frequency region characterized by intramolecular vibrations. (Adapted with permission from[77].)

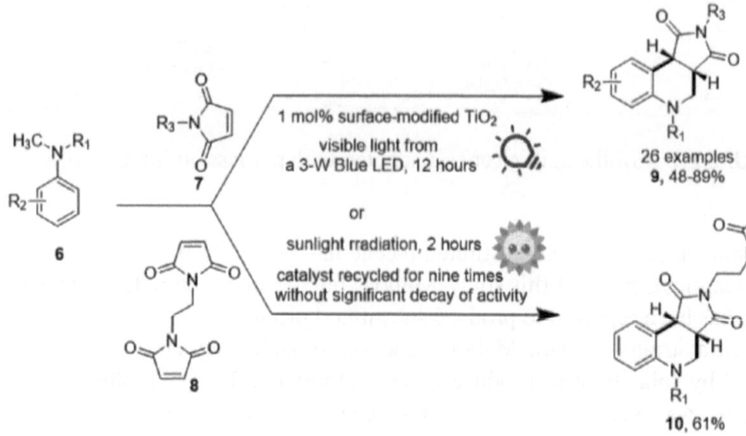

FIGURE 2.4   Cyclization of tertiary anilines with maleimides. (Adapted with permission from[78]. Copyright {2015} American Chemical Society.)

can be directly cyclized by visible light with maleimides to yield moderate to high yields of tetrahydroquinoline products at ambient temperatures (**Figure 2.4**)[78]. In contrast to unmodified TiO_2 catalysts, which are usually used in stoichiometric quantities in combination with UV light in conventional methods, a small amount of surface-modified TiO_2 catalyst combined with visible light is sufficient to catalyze the reaction efficiently. In comparison to transition metal complexes such as $Ru(bpy)_3Cl_2$ and $Ir(ppy)_2(dtbbpy)$ PF_6, surface-modified TiO_2 has a number of advantages, including low cost, high catalytic activity, easy recovery, and the ability to reuse nine times without significant catalytic degradation. Our exploratory research led us to become interested in using special properties of metals for a variety of purposes. Tellurium is an example of one of these metals[79-90].

When diisopropylethylamine (iPr2NEt) was present in acetonitrile (MeCN) with sunlight acting as a light source, the radical cyclization of alkyl bromide **11** and yield **12** was extremely efficient, as a result of a study using dimeric gold catalysts in combination with trialkyl amine bases to sacrificially donate electrons and hydrogen (**Figure 2.5**)[91].

In order to investigate the scope of this reaction, several alkyl or aryl bromide substrates were examined (**Tables 2.1 and 2.2**). The radical cyclization of alkyl bromides **11a–11c** with iPr2NEt in acetonitrile yielded cyclic products **12a–12c** with yields ranging from 58 to 92% (entries 1–6, **Table 2.1**). The yield of the reaction improved slightly when the reaction was irradiated with sunlight rather than UV light (entries 2, 4, and 6, **Table 2.1**).

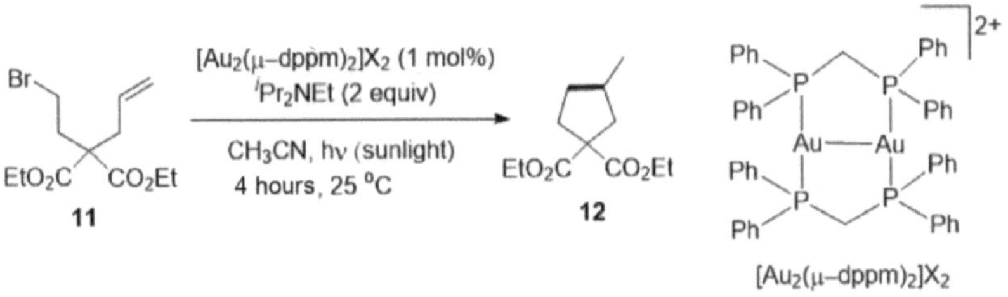

FIGURE 2.5 Optimization of the photoredox reaction. (Adapted with permission from[91].)

TABLE 2.1 Photoredox cyclization of alkyl bromides

ENTRY	SUBSTRATE	PRODUCT	PROCEDURE[a] (TYPE OF LIGHT)	YIELD [%] (RATIO)[b]
1 2	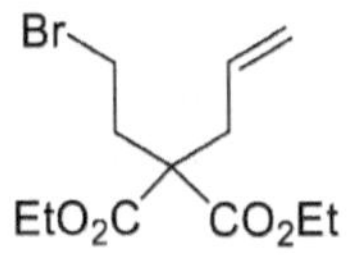 Br EtO_2C CO_2Et **11a**	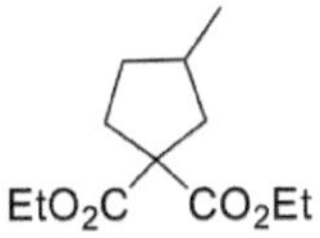 EtO_2C CO_2Et **12a**	A (UVA) A (sunlight)	81 (73:27) 92 (65:35)
3 4	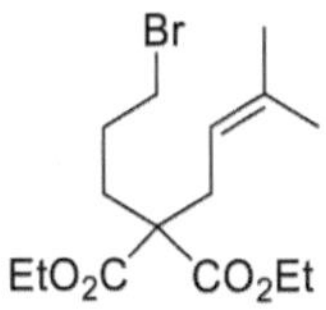 Br EtO_2C CO_2Et **11b**	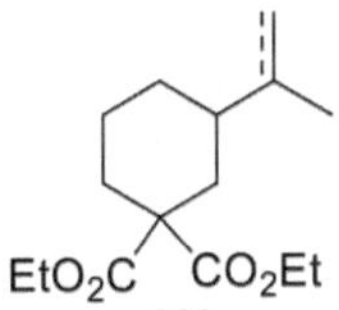 EtO_2C CO_2Et **12b**	A (UVA) A (sunlight)	58 (84:16) 81 (86:14)
5 6	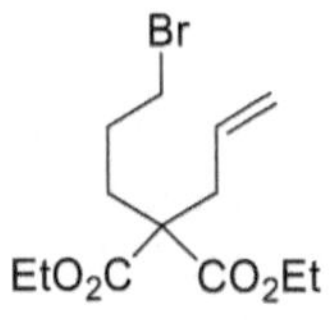 Br EtO_2C CO_2Et **11c**	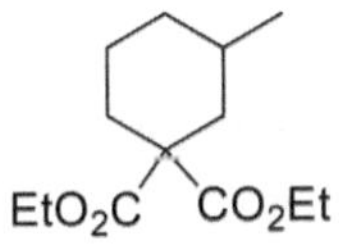 EtO_2C CO_2Et **12c**	A (UVA) A (sunlight)	58 60

Source: Adapted with permission from[91].

Note:
[a] Procedure A: [Au_2(μ-dppm)$_2$]Cl$_2$, iPr_2NEt, MeCN, 2–8 hours.
[b] Ratio of the isopropyl- to the isopropenyl-substituted product.

TABLE 2.2   Photoredox cyclization of aryl bromides

ENTRY	SUBSTRATE	PRODUCT	PROCEDURE[a] (TYPE OF LIGHT)	YIELD [%] (RATIO)
1	11d	12d	A (sunlight)	74 (71:29)
2 3	11e	12e	A (UVA) A (sunlight)	75[b] 80
4 5	11f	12f	A (UVA) A (sunlight)	90[b] 91

Source: Adapted with permission from[91]

Note:

[a] Procedure A: [Au$_2$(μ-dppm)$_2$]Cl$_2$, *i*Pr$_2$NEt, MeCN, 2–8 hours.

[b] For 8–16 hours, the reaction mixture was irradiated.

Due to the success of the dimeric gold complexes in photoredox reactions with alkyl bromides, aryl bromides were investigated for reductive cleavage (**Table 2.2**). By reductive radical cyclization, aryl bromide **11d** was converted into cyclized product **12d** containing isopropenyl- and isopropyl-substituted compounds in a 74% yield (entry 1, **Table 2.2**). Besides **12e** and **12f**, a number of other biaryl compounds were also prepared with excellent yields (entries 2–5, **Table 2.2**).

It is possible to activate and functionalize inert carbon-hydrogen bonds using visible light in combination with photoredox catalysis[92]. Several scientists have been interested in the functionalization of the sp3 carbon-hydrogen bond adjacent to a tertiary nitrogen atom[93]. Photoredox reactions mediated by visible light are catalyzed by iridium and ruthenium complexes as well as organic dyes[94]. As homogeneous catalysts, transition-metal complexes and organic dyes have many advantages, but they are expensive, difficult to recover, and cannot be recycled.

2.3 SOLAR LIGHT-MEDIATED THREE-COMPONENT REACTIONS

Pyrimidine compounds are among the many structural cores found in chemistry that have medical applications due to their antihypertensive, anticancer, anti-inflammatory, antiviral, and antibacterial properties. Many multicomponent methods exist for synthesizing pyriidines, but many of these do not work due to the high temperatures, long reaction times, and complex procedures involved. The use of a multicomponent process to combine aldehydes with thiourea or urea in order to get cyanopyrimidine derivatives was investigated in order to overcome some of these issues. It has also been reported that several other authors have used the Biginelli multicomponent reaction (MCR) protocol to produce diverse 1,2,3,4-tetrahydropyrimidines.

Due to their non-toxicity, thermal stability, chemical stability, optical stability, and strong oxidation activity, TiO_2 NPs have been used for the synthesis of several organic compounds. The main advantage of these materials lies in their high photocatalytic activity, which comes from their exposure to UV irradiation that generates electron-hole pairs. In the process of decomposing adsorbed matter, electrons and holes are diffused to the surface of reacted materials, resulting in efficient and environmentally friendly decomposition. TiO_2 has many advantages, which make it useful in organic synthesis processes, including esterification, hydrogenation, deoxygenation, water-gas shift reactions, and light-induced molecular transformations. Octanal-alkylation can also be activated by visible light using a dual catalyst based on TiO_2 and an organocatalyst (such as imidazolidinone).

A heterogeneous catalyst is a chemical catalyst whose physical phase differs from that of the products and/or reactants involved in the catalyzed reaction. Solid phase heterogeneous catalysts are typically used to facilitate reactions between two reactants. Combining light with a solid catalyst, however, can cause chemical reactions with heterogeneous PCs. Under benign conditions, heterogeneous photocatalysis can purify water and synthesize organic molecules, making it a promising technology for environmental conservation.

Three-component 1,2,3,4-tetrahydropyrimidine-5-carbonitrile derivatives have been prepared with photocatalysis[44]. Photocatalytic TiO_2 NPs (PCs) produced high yields under UV-visible illumination, as demonstrated by comparison with traditional thermal reactions under similar conditions. Using TiO_2 nanoparticles as organocatalysts and PCs, various derivatives of 1,2,3,4-tetrahydropyrimidine-5-carbonitrile were synthesized. According to the results of the study, TiO_2 showed excellent performance as a PC in terms of reaction time and yield. It is possible to adjust the amount of PC as well as the solvent system to optimize the photocatalytic activity. Since only a small amount of TiO_2 nanocatalyst is needed to complete the reaction when exposed to solar radiation, the proposed method offers a superior yield over conventional thermal method. As a result of analyzing pristine and recycled TiO2 PCs, it has been demonstrated that they are capable of being reused up to four times without losing any activity. In the past few years, we have made significant advances in various catalytic procedures. Our study has also shown that photo radiation can have a powerful effect on synthesis[3,4,95–99].

2.4 SOLAR LIGHT-INDUCED SYNTHESIS OF AMINOALKYL SYSTEMS

The hydroaminomethylation of olefins using aminomethyltrifluoroborate was catalyzed by photoredox catalysis[100]. An electron-deficient C=C bond can be introduced with a primary aminomethyl group using this methodology. It is possible to synthesize baclofen by this reaction, which is a derivative of aminobutyric acid (GABA).

Figure 2.6 shows a schematic representation of the optimized photocatalytic aminomethylation of methyl acrylate (**14**) using potassium Boc-protected aminomethyltrifluoroborate (**13**). Based on experimental evidence, **Figure 2.7** shows a plausible mechanism of reaction.

Numerous biologically active compounds possess the aminoalkyl structure[101–103]. Primary aminoalkyl motifs are found in many amino acids. In order to obtain aminoalkyl derivatives and pharmaceuticals containing aminoalkyl groups, a simple and selective procedure must be used for constructing the primary aminoalkyl component. Iridium cyclometalated complexes and ruthenium polypyridine derivatives have been proven to be

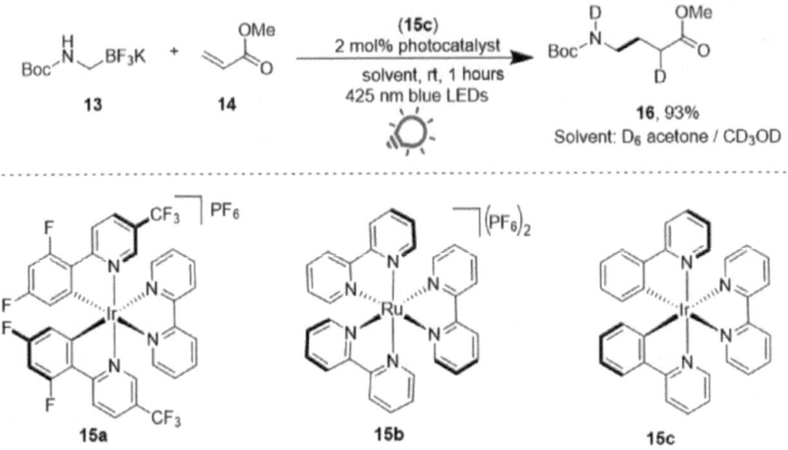

FIGURE 2.6 Optimization of the photocatalytic aminomethylation of methyl acrylate with potassium Boc-protected aminomethyltrifluoroborate. (Adapted with permission from[100].)

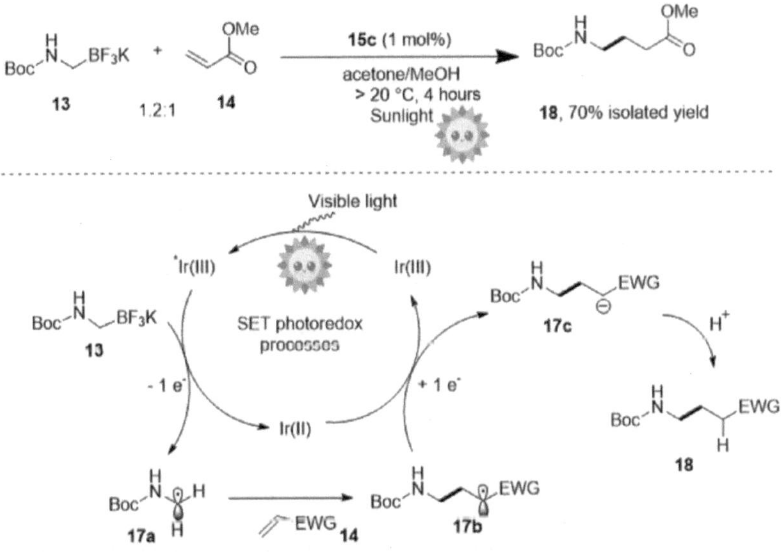

FIGURE 2.7 A plausible reaction mechanism. (Adapted with permission from[100].)

effective in oxidatively transforming electron-rich amines such as tertiary amines and enamines under visible light irradiation for many years[104–108]. During photoredox-catalyzed SET processes, reactive chemical species, such as iminium ions, azomethine ylides, and radicals, are generated, resulting in a variety of N-containing compounds. In previous studies, secondary and tertiary amines have mainly been presented, but primary amine preparation has not been discussed[109,110].

Because organoborates and organoboronic acid derivatives are stable and low toxic, as well as being compatible with a variety of functional groups, they are considered convenient radical precursors. The one-electron oxidation of organoboron compounds generates carbon-centered radicals during deboronation. Many metal-catalyzed radical reactions have used organoboronic acid derivatives, but the reaction often requires excess cooxidants (the upper process in **Figure 2.8**). Researchers have developed radical reactions of

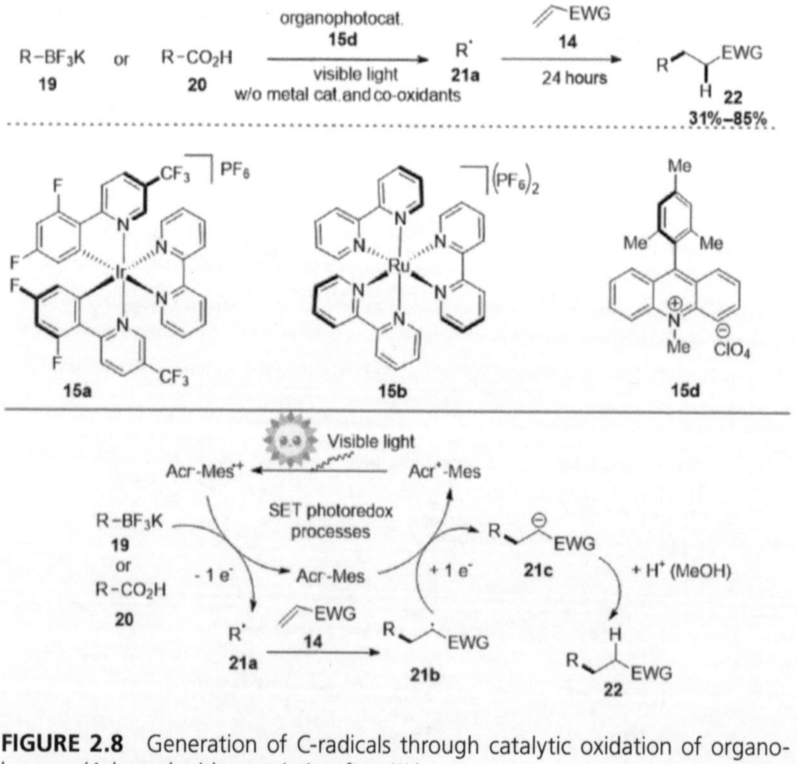

FIGURE 2.8 Generation of C-radicals through catalytic oxidation of organo-borons. (Adapted with permission from[111].)

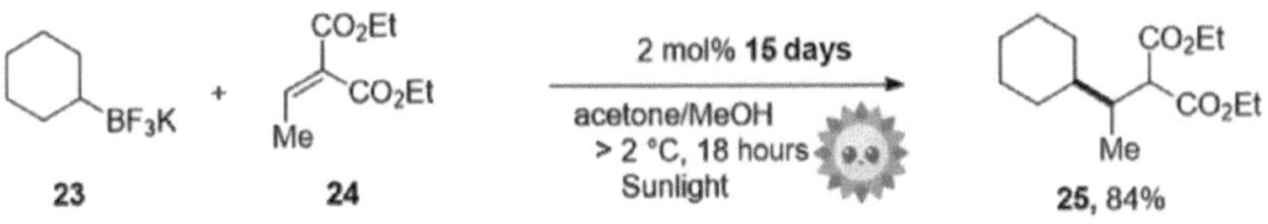

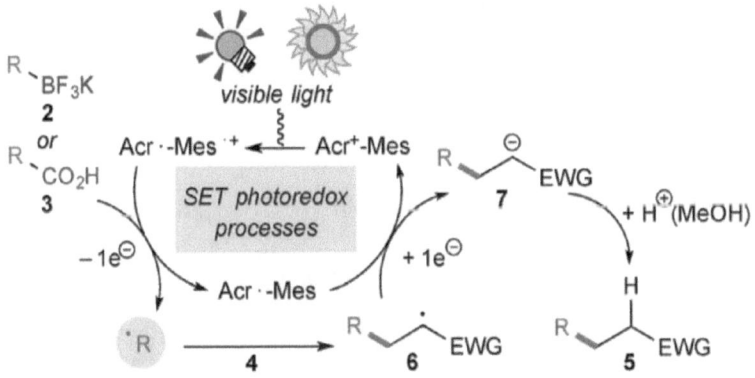

FIGURE 2.9   Reaction under sunlight. (Adapted with permission from[111].)

organoborates catalyzed by Ir or Ru photoredox catalysis (the lower process in **Figure 2.8**).

There have been reports in the literature of the creation of redox-economic radicals from carboxylic acids and organoborates through organic photoredox catalysis[111]. Carbon radicals can be generated by oxidizing carboxylic acids and organotrifluoroborates using an organophotoredox catalyst, 9-mesityl-10-methylacridinium perchlorate. The radical C–C bonds can be formed with electron-deficient olefins using this organophotocatalytic protocol. Using sunlight, which contains visible light, as a source of light is an extraordinary discovery. There was more efficiency with the sunlight-promoted process than with the blue LED system. It took 18 hours for the reaction between 2f and 4b to complete under daylight conditions (**Figure 2.9**).

A plausible mechanism involving redox-neutral processes is presented in **Figure 2.10**. The organocatalyst undergoes photoinduced electron transfer when visible light (sunlight blue or LEDs) is irradiated on it. Through the excited state, this serves as a strong oxidant. A carbon-centered radical is formed via the 1e-oxidation of carboxylic acid or organoborate by

FIGURE 2.10   A plausible reaction mechanism. (Adapted with permission from[111].)

Acr·–Mes·+ accompanying the creation of the reduced species, Acr·–Mes. A radical intermediate is formed when an organic radical reacts with an electron-defficient alkene. Also, a subsequent 1e-reduction process produced a carbanion intermediate. The adduct was then obtained by smooth protonation by the solvent, MeOH.

Currently, site-selective functionalization of alkanes is one of the most challenging aspects of organic chemistry. The ability of aliphatic nitriles to be converted into numerous functional compounds, including aldehydes, carboxylic acids, amides, esters, and amines, makes them potential compounds for use in a number of applications. The range of nitriles available would be significantly expanded by functionalizing the C–H bonds of aliphatic nitriles. A site-selective conversion of C/H/C cannot occur anywhere else except at alpha position in aliphatic nitriles. As a PC, tetrabutylammonium decatungstate (TBADT) was useful for regioselectively functionalizing cyclopentanone at the beta position with a moderate degree of selectivity for C–H acylation and alkylation. A polar effect of the excited decatungstate anion is invoked in order to explain why the normally activated alpha positions of the SH_2 transition state of hydrogen abstraction are inactive.

There has been a report on the conversion of aliphatic nitriles by photocatalysis[112]. A site-selective alkylation of aliphatic nitriles by alkenes was reported to be highly effective using photocatalysis. Using a decatungstate salt as the PC, the β- or γ-site-selective C–H alkylation of aliphatic nitriles was achieved. The observed site selectivity can be explained by a radical polar effect in the transition states for hydrogen abstraction. As shown in **Figure 2.11**, aliphatic nitriles are alkylated site selectively by TBADT PCs. Alkylation of aliphatic nitrile directly by sunlight has also been successful and produces the γ-alkylated nitrile in an adequate yield. In **Figure 2.12**, a plausible mechanism for the reaction is depicted.

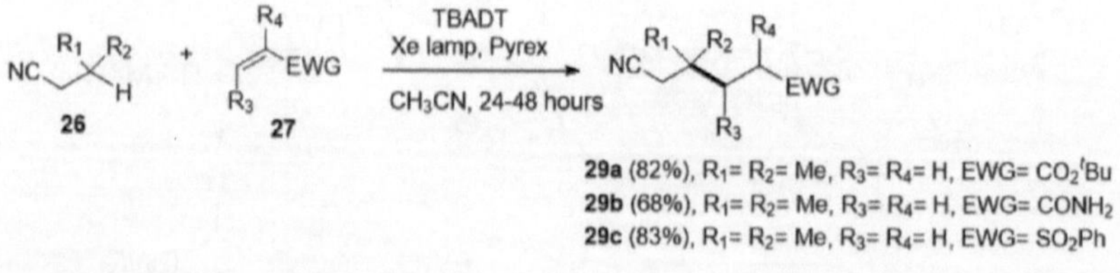

R1
R2
NC
H
26
+
R4
EWG
R3
27
TBADT
Xe lamp, Pyrex
CH3CN, 24-48 hours
R1
R2
R4
NC
EWG
R3
29a (82%), R1= R2= Me, R3= R4= H, EWG= CO2tBu
29b (68%), R1= R2= Me, R3= R4= H, EWG= CONH2
29c (83%), R1= R2= Me, R3= R4= H, EWG= SO2Ph

FIGURE 2.11 Using TBADT photocatalyst, the site-selective C–H alkylation by aliphatic nitriles 1. (Adapted with permission from[112]. Copyright {2015} American Chemical Society.)

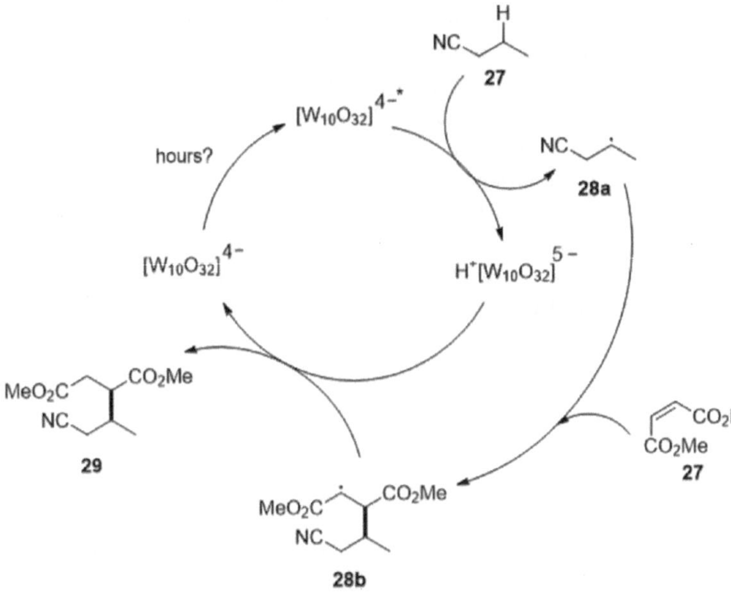

FIGURE 2.12 Proposed reaction mechanism for the photocatalyzed β–C–H alkylation of butyronitrile 1a with dimethyl maleate. (Adapted with permission from[112]. Copyright {2015} American Chemical Society.)

2.5 SOLAR LIGHT-INDUCED ADDITION REACTIONS ONTO IMINES AND OR DIAZO BONDS

As C–N bonds are extremely common in both natural and non-natural biologically active molecules, convenient methods for constructing them have been developed. The chemistry of azodicarboxylates has generated considerable interest in recent years. Diakyl azodicarboxylates offer a number of advantages as a synthetic tool, including their role in Mitsunobu reactions and their high electrophilicity demonstrated by Huisgen and his colleagues. As a result of pericyclic reactions such as hetero Diels-Alder reactions and azaene reactions, where the azo ester acts as both an enophile and a dienophile, valuable C–N bonds can be formed. The electrophilic properties of azodicarboxylates have also been explored in reactions with many nucleophiles, including enolates.

Using THF, cyclohexane, and heptanal as model substrates, the photoaddition of C–H bonds to diisopropyl azodicarboxylate (DIAD) was studied to understand how the reaction conditions affect the addition of C–H bonds to the N=N double bond (**Figure 2.13**). **33a** was produced in a 42% yield using a SolarBox equipped with a xenon lamp. THF photoaddition to compound **31** took place faster in the absence of TBADT, resulting in moderate yields of product **33b** after 2 hours, while in the presence of TBADT produced good yields of product **33b** after 2 hours. SolarBox experiments conducted using a similar reaction also went well and yielded more **33b**. The addition of heptanal (**30c**) to DIAD **31** utilizing a xenon lamp was slow, and only a very low

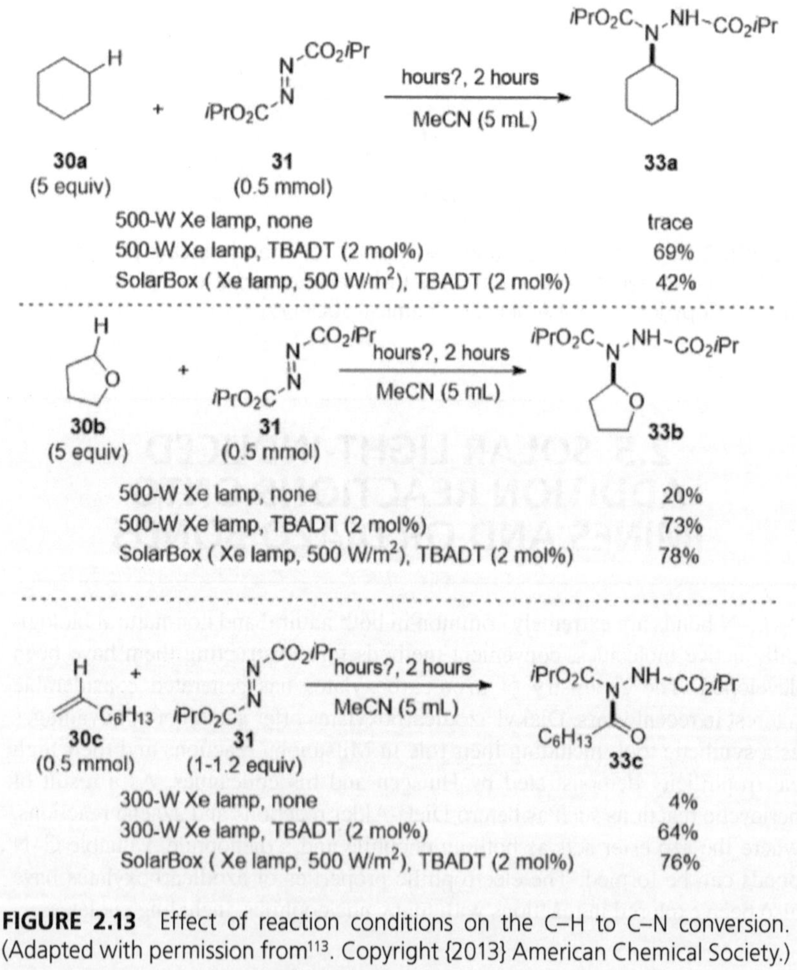

FIGURE 2.13 Effect of reaction conditions on the C–H to C–N conversion. (Adapted with permission from[113]. Copyright {2013} American Chemical Society.)

yield of **33c** was obtained even after 20 hours of reaction time. In the presence of 2 mol% of TBADT, the reaction accelerated dramatically. As a result, the amide was formed after 20 hours in a 69% yield and after 2 hours in a 64% yield. Using the SolarBox, a similar reaction resulted in the formation of the produce with a 76% yield.

There has been a study published describing the efficient conversion of C–H/C–N and C–H/C–CO–N via decatungstate photoinduced alkylation of DIAD[113]. Under conditions of irradiation, TBADT accelerated the addition of C–H bonds to the N=N double bond of DIAD. To generate the desired acyl hydrazides, three-component couplings induced by light were also achieved between cyclic alkanes, DIAD, and CO.

The catalytic conversion of C–H to C–CO–N (or C–N) may be based on a possible reaction mechanism (**Figure 2.14**). As a result, the excited polyoxodecatungstate anion abstracts a hydrogen atom from bond 1 and produces radical A (path a). A radical A undergoes consecutive additions to CO and DIAD **31** to form acyl radical B (path b) and aminyl radical C from it (path c). Through a back-hydrogen atom transfer from the reduced decatungstate anion to C, acyl hydrazides **33c** are formed, restoring the original TBADT (path d). As a result

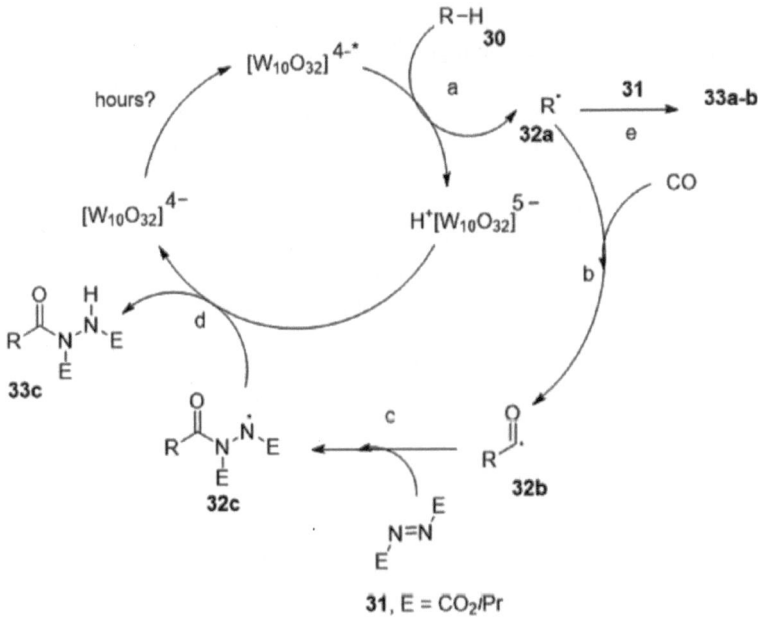

FIGURE 2.14 Possible mechanism for TBADT-photoinduced synthesis of 3. (Adapted with permission from[113]. Copyright {2013} American Chemical Society.)

of the absence of CO, hydrazides **33a-b** were produced smoothly (path e). Decarbonylation of acyl radicals generated from aldehydes was inhibited by the great capability of **31** to trap radicals completely or partially.

TBADT was found to enhance the light-promoted additions of diverse C–H bonds to DIAD in order to create C–N bonds as a result of the research. A three-component coupling reaction between DIAD, cyclic alkanes, and CO under pressure yielded acyl hydrazides, which may be useful as antioxidants. Additionally, in some instances, the greenness of the reaction can be observed by performing the process under solar illumination.

2.6 SOLAR LIGHT-MEDIATED 2- AND 3-FUNCTIONALIZED CARBONYL SYSTEMS

The regioselective beta-alkylation and acylation of cyclopentanones can be achieved using sunlight photocatalysis[114]. TBADT has been shown to induce direct, regioselective beta-alkylation of cyclopentanones with electron-deficient alkenes under bright sunlight. A polar transition state for an SH_2 reaction can be used to rationalize the regiochemistry. A good yield of the three-component beta-acylation product was obtained after CO was added to the reaction. Various electron-deficient alkene radical acceptors were screened, including methyl vinyl ketone (**35b**), acrylonitrile (**35a**), ethyl acrylate (**35c**), and cyclopentanone (**34**). In the process of converting beta-C–H/C–C, all of these reactions led to an exclusive selectivity (**Figure 2.15**).

In fact, the beta-acylation of cyclopentanone under a pressurized CO atmosphere has gone well. In this reaction, beta-keto radicals were formed,

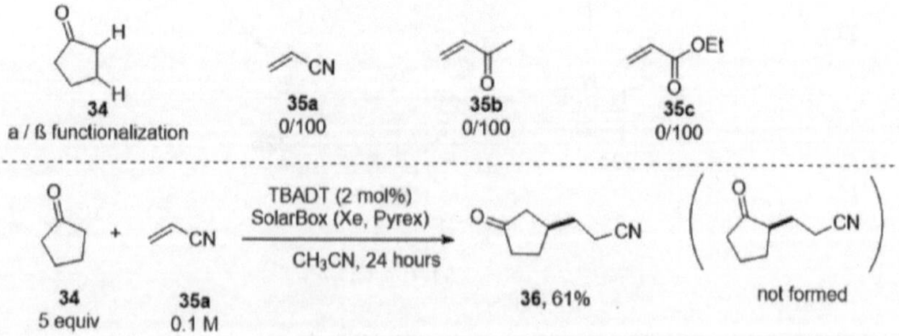

FIGURE 2.15 Regioselectivity in the TBADT-photocatalyzed C–H/C–C conversion in cyclopentanone (1a). (Adapted with permission from[114].)

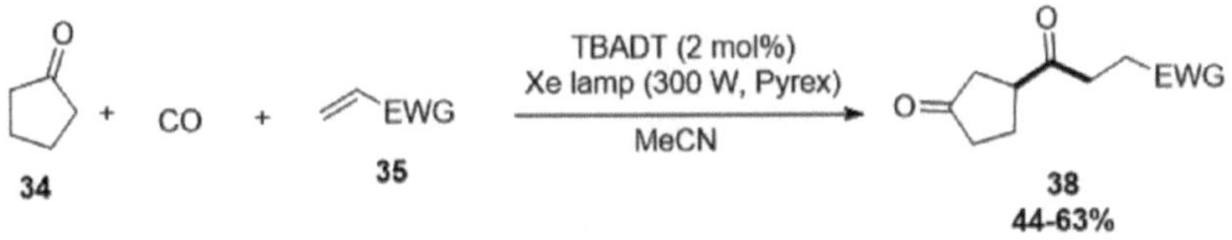

FIGURE 2.16 Three-component b-acylation of cyclopentanone (1a) using TBADT as the photocatalyst. (Adapted with permission from[114].)

which were sequentially trapped by electron-deficient alkenes and CO (**Figure 2.16**).

Figure 2.17 illustrates a plausible mechanism for the photocatalyzed beta-acylation of **34**. it was expected that excited polyoxotungstate anion $[W_{10}O_{32}]^{4-*}$ would have an electronegative oxygen character as the reactive site, and therefore the observed selectivity is in favor of beta-C–H abstraction (at least for **34**). Upon forming the beta-keto radical, it is successively added to

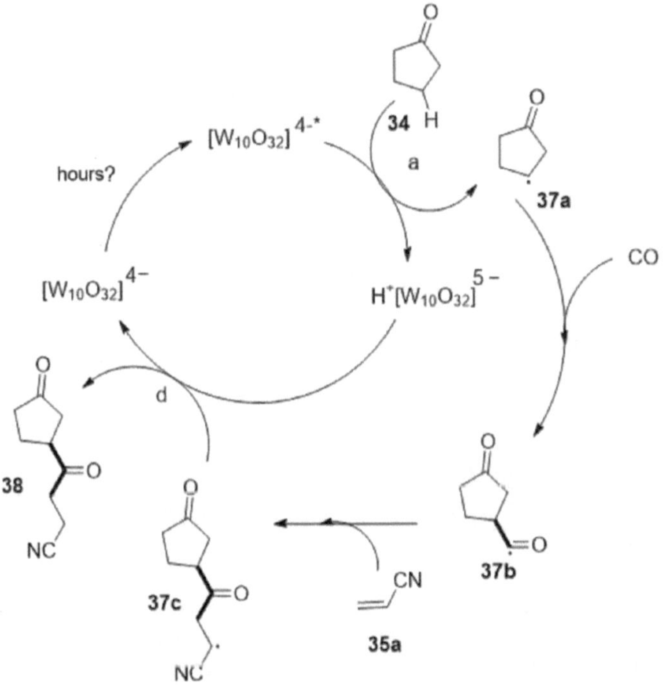

FIGURE 2.17 Proposed mechanism for the photocatalyzed b-C–H/C–C conversion in cyclopentanone and ensuing carbonylation. (Adapted with permission from[114].)

CO (if present) and to the electron-deficient olefin to form an adduct radical. A back-hydrogen atom transfer from the reduced form of the tungstate anion to the latter radical produces product **38**, restoring the TBADT catalyst.

Due to their extraordinary potential, sunlight-driven photochemical processes using asymmetric catalytic systems are exceptionally promising for synthesis of chiral molecules. Despite the short-lived electronic excited states inherent to any photochemical reaction, chiral catalysts are able to dictate the stereochemistry of the products, which can be a challenge. Through the photochemical activity of a key donor-acceptor complex, aldehydes can undergo stereoselective alpha-alkylation via catalytic activity[115]. In addition to being useful in thermal asymmetric processes, organic chiral catalysts are also capable of influencing the stereochemistry of carbon–carbon bond formation in visible light-driven reactions.

An innovative mechanism of catalysis was proposed, where the catalyst plays an active role in both the photochemical activation of the substrate and the stereoselectivity-defining propulsion of the reaction. Through this approach, it is possible to perform transformations that are extremely difficult to accomplish under thermal conditions, such as the formation of all-carbon quaternary stereocenters, asymmetric alkylation of aldehydes with alkyl halides, and remote stereochemistry control.

The site-selective conversion of sp_3 C–H bonds into C–C bonds is still a challenge in synthetic organic chemistry[116]. There has been considerable interest in both transition metal-catalyzed[117] and radical approaches to achieve this goal[118(p4)]. MacMillan and his colleagues have developed a unique method for selectively converting b-C–H bonds rather than a-C–H bonds in ketones. An organocatalytic method is available for generating radicals from enamines generated by an organocatalytic reaction between a ketone and an amine[119]. Furthermore, the ketone can be converted in situ into its unsaturated derivative and then functionalized[120].

2.7 SOLAR LIGHT-INDUCED CHIRAL OXIDATION

An article has been published that describes highly active catalysts for aerobic oxidation driven by visible light based on bimetallic alloy nanoparticles supported on anatase TiO_2[121]. The irradiation of visible light on nanoparticles of Pt–Cu bimetallic alloy supported on anatase TiO2 promotes aerobic oxidation. As a result of visible light's interband excitation of Pt atoms, this process is facilitated, and the activated electrons are then transferred to the conduction band of anatase. During the photocatalytic process, substrates are oxidized due to the positive charges on nanoparticles, and electrons in the conduction band reduce molecular oxygen.

With chiral phase-transfer catalysis, enantioselective photooxygenation of beta-keto esters has been reported using molecular oxygen during the reaction[122]. In the presence of chiral phase-transfer catalysts, enantioselective photo-oxygenation of beta-keto esters with molecular oxygen has been achieved using a highly efficient and green method that involves catalyzing the reaction. In this reaction, high yields and good enantiomeric excesses have been demonstrated, and a plausible reaction mechanism has been proposed to explain the results.

A selective photocatalytic oxidation of benzyl alcohol and its derivatives into corresponding aldehydes by molecular oxygen on TiO_2 under visible light irradiation has been reported in the literature[123]. A TiO2 PC under an atmosphere of oxygen was used to carry out the photocatalytic oxidation of benzyl alcohol and its derivatives, such as 4-chlorobenzyl alcohol, 4-methoxybenzyl alcohol, 4-nitrobenzyl alcohol, 4-(trifluoromethyl)benzyl alcohol, 4-methylbenzyl alcohol, and 4-tertiarybutylbenzyl alcohol, into corresponding aldehydes. To verify that the reaction would proceed, visible light and UV light were used to irradiate the surface. Particularly, the visible light response was observed to be attributed to the formation of a characteristic surface complex formed by the adsorption of benzyl alcoholic compounds onto the surface of TiO2. In addition to studying the properties of the surface complex as the active center for this selective photocatalytic oxidation, the mechanism behind it has also been studied.

According to the results obtained on the Pt–Cu alloy catalyst, the apparent quantum yield for the reaction is about 17% when the reaction is exposed to monochromatic 550 nm light, which is significantly higher than the yield obtained on the monometallic Pt catalyst. The work function of nanoparticles is reduced when Cu is alloyed with Pt, and the Schottky barrier formed at the interface between nanoparticles and anatase is also reduced. Thus, photoactivated nanoparticles and anatase are able to transfer electrons efficiently, enhancing photocatalytic activity. In addition to being able to be activated by sunlight, the Pt-Cu alloy catalyst is capable of oxidizing alcohols efficiently and selectively at ambient temperatures.

2.8 SOLAR LIGHT-MEDIATED EFFICIENT ECO-FRIENDLY DEGRADATION

Along with industrial development around the world, excessive release of antibiotic residues into the aquatic environment has posed a serious threat to the life of living ecosystems. As one of the most widely used antibiotics in the world, tetracycline (TC) plays an important role in the prevention and treatment of bacterial infections in humans and animals[124]. In spite of this, the removal of TC via conventional wastewater treatment faces a number of challenges due to TC's incomplete metabolization in the body and in the environment due to its low biodegradability. Only a few reports have attempted to use direct sunlight for the degradation of TC through a photocatalytic pathway[125,126]. Several of these methods have a number of drawbacks, including the need for large amounts of PC, the low degradation efficiency, incomplete PC recovery, prolonged time periods, and the need for expensive or complex photocatalytic systems. It is necessary to develop a method of dealing with TC contamination that is more efficient, sustainable, and convenient.

Using ZnO/NiFe$_2$O$_4$/Co$_3$O$_4$ as a highly efficient magnetically separable PC, a solar light-induced photocatalytic degradation of TC has been demonstrated[127].

Photocatalytic degradation of TC under natural sunlight irradiation showed satisfactory PC activity (**Figures 2.18 and 2.19**). The optimal degradation amount for TC in the presence of ZnO/NiFe$_2$O$_4$/Co$_3$O$_4$ in pH of 9, under the solar light illumination, has been found to be 98% within 20 minutes of reaction time. Moreover, it is noteworthy that the superb

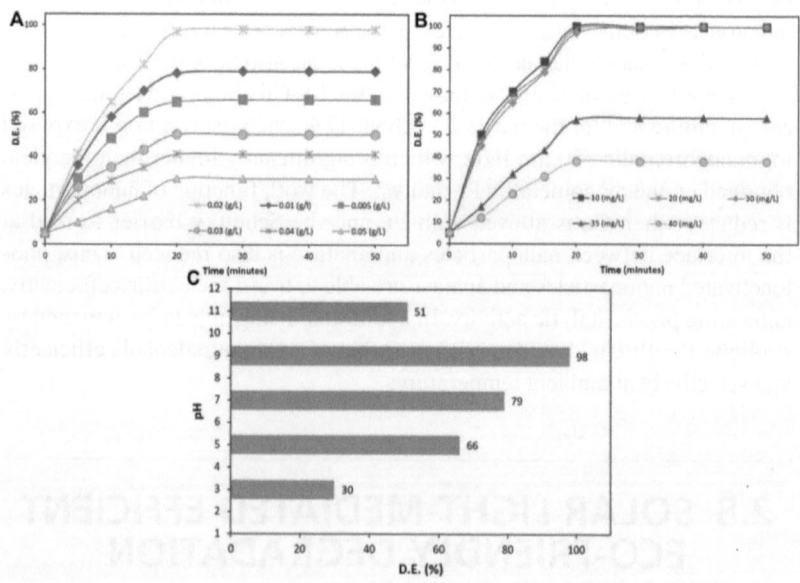

FIGURE 2.18 (A) The effect of photocatalyst amount on the efficiency of the TC photodegradation, (B) TC concentration effect on the progress of the photodegradation process, and (C) the effect of pH on TC degradation. (Adapted from[127].)

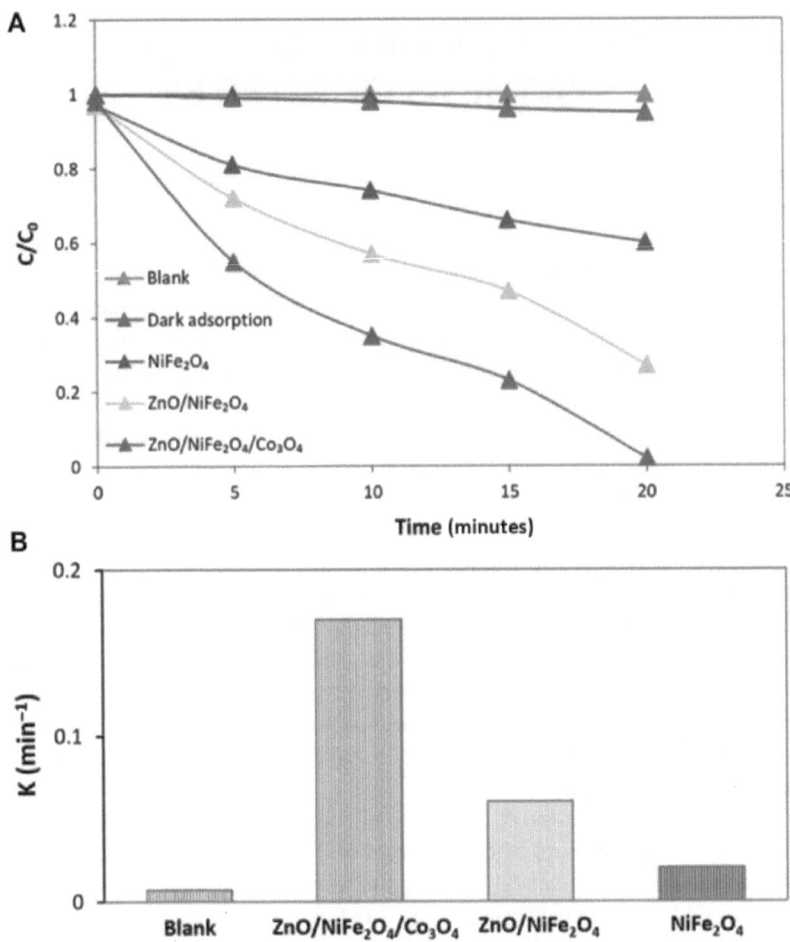

FIGURE 2.19 (A) Sunlight driven photocatalytic degradation of TC under optimal conditions and (B) the degradation rate constants for the sunlight-driven photocatalytic degradation of TC under optimal conditions. (Adapted with permission from[127].)

activity of the $ZnO/NiFe_2O_4/Co_3O_4$ may be due to the synergistic optical effects of ZnO, $NiFe_2O_4$, and Co_3O_4 that result in the substantial separation of charge carriers, which has a direct impact on reducing the recombination speed of the electrons/holes created by the photogeneration process. In no doubt, this study could provide inspiration for the sustainable remediation of a wide range of contaminants.

2.9 SOLAR LIGHT-INDUCED SYNTHESIS OF NITRILES

It is important to note that nitriles are one of the most important compounds used as intermediates for the manufacture of agricultural chemicals, dyes, polymers, and fine chemicals[128]. Furthermore, compounds that contain cyano groups can be used as pharmaceuticals and functional materials[129,130]. Traditional nitrile syntheses, such as the Sandmeyer reaction or the ammoxidation, produce excessive amounts of waste or require high temperatures[131]. Several alternative pathways are currently being investigated, including amide transformations[132], aldoxime transformations[133], alcohol transformations[134], and amine transformations[135]. At temperatures around 100°C, supported hydrous ruthenium oxide is effective for aerobically oxidizing amines to nitriles. Both ruthenium oxide and hydrous ruthenium oxide are also known for their metallic conductivity, so both of these oxides can be combined with semiconductors to form photocatalytic systems[136,137].

An aerobic oxidation of amines to nitriles under visible light conditions using hydrous ruthenium oxide supported on TiO_2 has been reported[138]. In the process of converting benzylic and aliphatic amines into their corresponding nitriles under the atmospheric pressure of O_2, both LEDs (blue, green, or red) as well as the Sun served as effective light sources. It is possible to use water as a solvent instead of toluene without sacrificing the activity and selectivity.

An aerobic oxidation of amines to nitriles under visible light conditions using hydrous ruthenium oxide supported on TiO_2 has been reported[138]. In the process of converting benzylic and aliphatic amines into their corresponding nitriles under the atmospheric pressure of O_2, both LEDs (blue, green, or red) as well as the Sun served as effective light sources. Without sacrificing activity or selectivity, water can be used instead of toluene as a solvent.

2.10 CONCLUSIONS

Sunlight provides abundant and readily available energy that can be used to drive chemical reactions, which is an important step toward achieving a sustainable future. The mechanism of the reactions offers new insights of these processes. With UV radiation only representing 5% of sunlight energy, developing catalytic systems powered by visible light is particularly interesting. Catalyzing numerous reactions, such as alcohol oxidation, nitro aromatic compound reduction, and Suzuki-Miyaura cross-coupling, have been accomplished using such systems in recent years.

ACKNOWLEDGMENTS

AD is grateful to CEA-Grenoble; Joseph Fourier University; University of Göttingen; University of California, Los Angeles; and Prince Mohammad Bin Fahd University for their support. BKB is grateful to US NIH, US NCI, Texas Kleberg Foundation, Stevens Institute of Technology, University of Texas M. D. Anderson Cancer Center, University of Texas-Pan American, Community Health Systems of Texas, and Prince Mohammad Bin Fahd University for their support.

REFERENCES

1. Das A. Recent developments in semipolar InGaN laser diodes. *Semiconductors.* 2021;55(2):272–282. doi:10.1134/S106378262102010X
2. Das A. A systematic exploration of InGaN/GaN quantum well-based light emitting diodes on semipolar orientations. *Opt Spectrosc.* 2022;130(3):137–149. doi:10.1134/S0030400X2203002X
3. Yadav RN, Hossain F, Das A, Srivastava AK, Banik BK. Organocatalysis: a recent development on stereoselective synthesis of o-glycosides. *Catal Rev.* 2022;66(1):1–118. doi:10.1080/01614940.2022.2041303
4. Das A, Yadav RN, Banik BK. A novel Baker's yeast-mediated microwave-induced reduction of racemic 3-keto-2-azetidinones: facile entry to optically active hydroxy β-lactam derivatives. *Curr Organocatalysis.* 2022;9(2):195–198.
5. Das A, Banik BK. Versatile synthesis of organic compounds derived from ascorbic acid. *Curr Organocatalysis.* 2022;9(1):14–33.
6. Das A, Banik BK. Dipole Moment Studies on Beta Lactams. In: Banik BK, ed. *Green Approaches in Medicinal Chemistry for Sustainable Drug Design.* Elsevier; 2024:523–542.
7. Das A, Banik BK. Dipole moment in medicinal research: green and sustainable approach. In: Banik BK, ed. *Green Approaches in Medicinal Chemistry for Sustainable Drug Design.* Elsevier; 2024:561–602.
8. Das A, Das A, Banik BK. Influence of dipole moments on the medicinal activities of diverse organic compounds. *J. Indian Chem Soc.* 2021;98(2):100005. doi:10.1016/j.jics.2021.100005

9. Das A, Banik BK. β-lactams: geometry, dipole moment and anticancer activity. *J. Indian Chem Soc*. 2020;97(11b, Nov 2020):2461–2467. doi:10.5281/zenodo.5656689

10. Das A, Alqashqari AA, Banik BK. Quantum mechanical calculations of dipole moment of diverse imines. *J Indian Chem Soc*. 2021;97(9b):1563–1566.

11. Das A, Banik BK. Dipole moment studies on α-hydroxy-β-lactam derivatives. *J Indian Chem Soc*. 2021;97(9b):1567–1571.

12. Das A, Banik BK. Dipole moment and anticancer activity of beta lactams. *Indian J Pharm Sci*. 2021;83(5):1071–1074. doi:10.36468/pharmaceutical-sciences.862

13. Das A, Banik BK. Computational studies of physicochemical parameters on optically active anticancer β-lactams. *Heterocycl Lett*. 2023;13(1):17–26.

14. Das A, Yadav R, Banik BK. Dipole moment studies on anticancer polyaromatic compounds. *Heterocycl Lett*. 2024;14(2):287–291.

15. Das A, Banik BK. Studies on dipole moment of penicillin isomers and related antibiotics. *J Indian Chem Soc*. 2020;97(6):911–915.

16. Das A, Bose AK, Banik BK. Stereoselective synthesis of β-lactams under diverse conditions: unprecedented observations. *J Indian Chem Soc*. 2020;97(6):917–925.

17. Das A, Banik BK. 26 - Dipole Moment in Medicinal Research: Green and Sustainable Approach. In: Banik BK, ed. *Green Approaches in Medicinal Chemistry for Sustainable Drug Design*. Advances in Green and Sustainable Chemistry. Elsevier; 2020:921–964. doi:10.1016/B978-0-12-817592-7.00021-6

18. Das A, Banik BK. *Microwaves in Chemistry Applications: Fundamentals, Methods and Future Trends*. Elsevier Science; 2021.

19. Das A, Banik BK. Chapter 1 - Foundational Principles of Microwave Chemistry. In: Das A, Banik B, eds. *Microwaves in Chemistry Applications*. Advances in Green and Sustainable Chemistry. Elsevier; 2021:3–26. doi:10.1016/B978-0-12-822895-1.00005-9

20. Das A, Banik BK. Chapter 2 - Microwave Equipment for Chemistry. In: Das A, Banik B, eds. *Microwaves in Chemistry Applications*. Advances in Green and Sustainable Chemistry. Elsevier; 2021:27–59. doi:10.1016/B978-0-12-822895-1.00002-3

21. Das A, Banik BK. Chapter 3 - Modeling and Interpreting Microwave Effects. In: Das A, Banik B, eds. *Microwaves in Chemistry Applications*. Advances in Green and Sustainable Chemistry. Elsevier; 2021:61–104. doi:10.1016/B978-0-12-822895-1.00007-2

22. Das A, Banik BK. Chapter 4 - Microwave-Assisted Synthesis of Oxygen- and sulfur-Containing Organic Compounds. In: Das A, Banik B, eds. *Microwaves in Chemistry Applications*. Advances in Green and Sustainable Chemistry. Elsevier; 2021:107–142. doi:10.1016/B978-0-12-822895-1.00010-2

23. Das A, Banik BK. Chapter 5 - Microwave-Assisted Synthesis of N-Heterocycles. In: Das A, Banik B, eds. *Microwaves in Chemistry Applications*. Advances in Green and Sustainable Chemistry. Elsevier; 2021:143–198. doi:10.1016/B978-0-12-822895-1.00006-0

24. Das A, Banik BK. Chapter 6 - Microwave-Assisted Oxidation and Reduction Reactions. In: Das A, Banik B, eds. *Microwaves in Chemistry Applications*. Advances in Green and Sustainable Chemistry. Elsevier; 2021:199–244. doi:10.1016/B978-0-12-822895-1.00001-1

25. Das A, Banik BK. Chapter 7 - Microwave-Assisted Enzymatic Reactions. In: Das A, Banik B, eds. *Microwaves in Chemistry Applications*. Advances in Green and Sustainable Chemistry. Elsevier; 2021:245–281. doi:10.1016/B978-0-12-822895-1.00009-6

26. Das A, Banik BK. Chapter 8 - Microwave-Assisted Sterilization. In: Das A, Banik B, eds. *Microwaves in Chemistry Applications*. Advances in Green and Sustainable Chemistry. Elsevier; 2021:285–328. doi:10.1016/B978-0-12-822895-1.00011-4

27. Das A, Banik BK. Chapter 9 - Microwave-Assisted CVD Processes for Diamond Synthesis. In: Das A, Banik B, eds. *Microwaves in Chemistry Applications*. Advances in Green and Sustainable Chemistry. Elsevier; 2021: 329–374. doi:10.1016/B978-0-12-822895-1.00004-7

28. Das A, Banik BK. Chapter 10 - Future Trends in Microwave Chemistry and Biology. In: Das A, Banik B, eds. *Microwaves in Chemistry Applications*. Advances in Green and Sustainable Chemistry. Elsevier; 2021:375–384. doi:10.1016/B978-0-12-822895-1.00003-5

29. Das A, Yadav RN, Banik BK. Microwave-induced conversion of electromagnetic energy into heat energy in different solvents: synthesis of beta-lactams. *Chem J Mold*. 2022;17(1):62–66. doi:10.19261/cjm.2021.864

30. Das A, Banik BK. Microwave-induced biocatalytic reactions toward medicinally important compounds. *Phys Sci Rev*. 2022;7(4–5):507–538. doi:10.1515/psr-2021-0064

31. Das A, Banik BK. 3 Microwave-Induced Biocatalytic Reactions Toward Medicinally Important Compounds. In: *3 Microwave-Induced Biocatalytic Reactions Toward Medicinally Important Compounds*. De Gruyter; 2022: 57–88. doi:10.1515/9783110732542-003

32. Das A, Yadav R, Banik B. Microwave-induced surface-mediated highly efficient regioselective nitration of aromatic compounds: effects of penetration depth. *Asian J Chem*. 2021;33:2203–2206. doi:10.14233/ajchem.2021.23131

33. Das A, Banik BK. Microwave-induced catalytic transfer hydrogenation in different solvents toward optically active hydroxy beta lactams: effects of penetration depth. Asian J Org Med Chem. 2023;8(1):7–10.

34. Das A, Banik BK. Microwave in research-more miracles. *Heterocycl Lett*. 2024;14(2):449–456.

35. Das A, Banik BK. Expeditious synthesis of oxygen and sulfur heterocycles by microwave. *Heterocycl Lett*. 2024;14(2):449–456.

36. Das A, Yadav R, Banik BK. Microwave-induced ferrier rearrangement of hyroxy beta-lactams with glycals. *Appl Chem Eng*. 2024;7(2):1870–1870.

37. Ravelli D, Protti S, Fagnoni M. Carbon–carbon bond forming reactions via photogenerated intermediates. *Chem. Rev.* 2016;116(17):9850–9913.

38. Lang X, Zhao J, Chen X. Cooperative photoredox catalysis. *Chem. Soc. Rev.* 2016;45(11):3026–3038.

39. Yoon TP. Photochemical stereocontrol using tandem photoredox–chiral Lewis acid catalysis. *Acc Chem Res*. 2016;49(10):2307–2315.

40. Chen JR, Hu XQ, Lu LQ, Xiao WJ. Visible light photoredox-controlled reactions of N-radicals and radical ions. *Chem Soc Rev*. 2016;45(8):2044–2056.

41. Shaw MH, Twilton J, MacMillan DW. Photoredox catalysis in organic chemistry. *J Org Chem*. 2016;81(16):6898–6926.

42. Fabry DC, Rueping M. Merging visible light photoredox catalysis with metal catalyzed C–H activations: on the role of oxygen and superoxide ions as oxidants. *Acc Chem Res.* 2016;49(9):1969–1979.

43. Das A, NareshYadav R, KrishnaBanik B. Microwave-induced ferrier rearrangement of hyroxy beta-lactams with glycals. *Appl Chem Eng.* 2024;7(2): 1870–1870. doi:10.59429/ace.v7i2.1870

44. Alharthi AA, Alotaibi M, Shalwi MN, Qahtan TF, Ali I, Alshehri F, et al. Photocatalytic-driven three-component synthesis of 1,2,3,4-tetrahydropyrimidine-5-carbonitrile derivatives: A comparative study of organocatalysts and photocatalysts. *J Photochem Photobiol A: Chemistry.* 2023;436:114358. doi:10. 1016/j.jphotochem.2022.114358

45. Yang MQ, Shen L, Lu Y, Chee SW, Lu X, Chi X, et al. Disorder engineering in monolayer nanosheets enabling photothermic catalysis for full solar spectrum (250-2500 nm) harvesting. *Angew Chem Int Ed Engl.* 2019;58(10):3077–3081. doi:10.1002/anie.201810694

46. Lim PF, Leong KH, Sim LC, Oh WD, Chin YH, Saravanan P, et al. Mechanism insight of dual synergistic effects of plasmonic Pd-SrTiO3 for enhanced solar energy photocatalysis. *Appl Phys A.* 2020;126(7):550. doi:10.1007/s00339-020-03739-4

47. Das A, Banik BK. 15 - Versatile Thiosugars in Medicinal Chemistry. In: Banik BK, ed. *Green Approaches in Medicinal Chemistry for Sustainable Drug Design.* Advances in Green and Sustainable Chemistry. Elsevier; 2020:549–574. doi:10.1016/B978-0-12-817592-7.00015-0

48. Das A, Banik BK. Versatile Thiosugars in Medicinal Chemistry. In: Banik BK, ed. *Green Approaches in Medicinal Chemistry for Sustainable Drug Design.* Elsevier; 2024:409–441.

49. Das A, Yadav RN, Banik BK. Ascorbic acid-mediated reactions in organic synthesis. *Curr. Organocatalysis.* 2020; 7(3):212–241.

50. Das A, Banik BK. Graphene Oxide and Modified Graphene Oxide-Mediated Synthesis of Medicinally Active Compounds. In: Banik BK, ed. *Green Approaches in Medicinal Chemistry for Sustainable Drug Design.* Elsevier; 2024:13–44.

51. Das A, Banik BK. Synthesis of Natural Products by Photochemistry. In: Banik BK, ed. *Green Approaches in Medicinal Chemistry for Sustainable Drug Design.* Elsevier; 2024:259–283.

52. Das A, Banik BK. Green Synthesis of Biologically Active pyrroles and related substrates via C-H Functionalization. In: Banik BK, ed. *Green Approaches in Medicinal Chemistry for Sustainable Drug Design.* Elsevier; 2024:101–131.

53. Das A, Ashraf MW, Banik BK. Thione derivatives as medicinally important compounds. *ChemistrySelect.* 2021;6(34):9069–9100. doi:10.1002/slct.202102398

54. Zhang Y, Qian R, Zheng X, Zeng Y, Sun J, Chen Y, et al. Visible light induced cyclopropanation of dibromomalonates with alkenes via double-SET by photoredox catalysis. *Chem Commun.* 2014;51(1):54–57. doi:10.1039/C4CC08203F

55. Silva MDL, David JP, Silva LCRC, Santos RAF, David JM, Lima LS, et al. Bioactive oleanane, lupane and ursane triterpene acid derivatives. *Molecules.* 2012;17(10):12197–12205. doi:10.3390/molecules171012197

56. Lin KW, Huang AM, Lin CC, Chang CC, Hsu WC, Hour TC, et al. Anti-cancer effects of ursane triterpenoid as a single agent and in combination with cisplatin in bladder cancer. *Eur J Pharmacol.* 2014;740:742–751. doi:10.1016/j.ejphar. 2014.05.051

57. Putta S, Yarla NS, Kilari EK, Surekha C, Aliev G, Divakara MB, et al. Therapeutic potentials of triterpenes in diabetes and its associated complications. *Curr Top Med Chem.* 2016;16(23):2532–2542.
58. Safayhi H, Sailer ER. Anti-inflammatory actions of pentacyclic triterpenes. *Planta Med.* 1997;63(06):487–493. doi:10.1055/s-2006-957748
59. Dey Ray S, Dewanjee S. Isolation of a new triterpene derivative and in vitro and in vivo anticancer activity of ethanolic extract from root bark of *Zizyphus nummularia* Aubrev. *Nat. Prod. Res.* 2015;29(16):1529–1536. doi:10.1080/1478 6419.2014.983921
60. Ray SD, Ray S, Zia-Ul-Haq M, De Feo V, Dewanjee S. Pharmacological basis of the use of the root bark of *Zizyphus nummularia* Aubrev. (Rhamnaceae) as anti-inflammatory agent. *BMC Complement Altern Med.* 2015;15(1):416. doi:10.1186/s12906-015-0942-7
61. Das A, Banik BK. Advances in heterocycles as DNA intercalating cancer drugs. *Phys Sci Rev.* 2023;8(9): 2473–2521. Published online January 5, 2022. doi:10.1515/psr-2021-0065
62. Das A, Banik BK. 4 Advances in Heterocycles as DNA Intercalating Cancer Drugs. In: *Heterocyclic Anticancer Agents.* De Gruyter; 2022:111–160. doi:10. 1515/9783110735772-004
63. Banik BK, Das A. *Natural Products as Anticancer Agents.* Elsevier Science; 2023.
64. Banik BK, Das A. Anticancer Activity of Natural Compounds from Marine Plants. In: Banik BK, Das A, eds. *Natural Products as Anticancer Agents.* Elsevier; 2024:237–284.
65. Banik BK, Das A. Anticancer Activity of Natural Compounds from Bacteria. In: Banik BK, Das A, eds. *Natural Products as Anticancer Agents.* Elsevier; 2024:287–328.
66. Banik BK, Das A. Anticancer Activity of Natural Compounds from Fungi. In: Banik BK, Das A, eds. *Natural Products as Anticancer Agents.* Elsevier; 2024:329–366.
67. Banik BK, Das A. Anticancer Drugs from Hormones and Vitamins. In: Banik BK, Das A, eds. *Natural Products as Anticancer Agents.* Elsevier; 2024:369–414.
68. Banik BK, Das A. Future Prospect in Anticancer Natural Products. In: Banik BK, Das A, eds. *Natural Products as Anticancer Agents.* Elsevier; 2024:415–426.
69. Das A, Banik BK. Anticancer Activity of Natural Compounds from Leaves of the Plants. In: Banik BK, Das A, eds. *Natural Products as Anticancer Agents.* Elsevier; 2024:3–48.
70. Das A, Banik BK. Anticancer Activity of Natural Compounds from stems/ barks of the Plants. In: Banik BK, Das A, eds. *Natural Products as Anticancer Agents.* Elsevier; 2024:49–86.
71. Das A, Banik BK. Anticancer Activity of Natural Compounds from Roots of the Plants. In: Banik BK, Das A, eds. *Natural Products as Anticancer Agents.* Elsevier; 2024:87–132.
72. Das A, Banik BK. Anticancer Activity of Natural Compounds from Fruits and Vegetables. In: Banik BK, Das A, eds. *Natural Products as Anticancer Agents.* Elsevier; 2024:133–178.

73. Das A, Banik BK. Anticancer Activity of Natural Compounds from Marine Animals. In: Banik BK, Das A, eds. *Natural Products as Anticancer Agents*. Elsevier; 2024:181–236.

74. Das A, Banik BK. Combatting the coronavirus utilizing natural cinnamon and its derived products. *Asian J Synth Nat Prod Chem*. 2023;1(1):11–15.

75. Das A. Quantitative structure-property relationships of taxol, taxotere and their epi-isomers. *J Indian Chem Soc*. 2020;97(11):9.

76. Shaikh AL, Das A, Banik BK. Indium-mediated reduction of aromatic nitro groups in β-lactams to oxazines. *Heterocycl Letters*. 2024;14(2):267–272.

77. Protti S, Artioli GA, Capitani F, Marini C, Dore P, Postorino P, et al. Preparation of (substituted) picenes via solar light-induced Mallory photocyclization. *RSC Adv*. 2015;5(35):27470–27475. doi:10.1039/C5RA02855H

78. Tang J, Grampp G, Liu Y, Wang BX, Tao FF, Wang LJ, et al. Visible light mediated cyclization of tertiary anilines with maleimides using nickel(ii) oxide surface-modified titanium dioxide catalyst. *J Org Chem*. 2015;80(5):2724–2732. doi:10.1021/jo502901h

79. Das A, Banik BK Tellurium-based solar cells. *Phys Sci Rev*. 2022;8(12):4631–4658. Published online May 18, 2022. doi:10.1515/psr-2021-0110

80. Das A, Banik BK. 5 Tellurium-Based Solar Cells. In: *5 Tellurium-Based Solar Cells*. De Gruyter; 2022:107–134. doi:10.1515/9783110735840-005

81. Das A, Banik BK. Semiconductor characteristics of tellurium and its implementations. *Phys Sci Rev*. 2022;8(12):4659–4687. Published online May 18, 2022. doi:10.1515/psr-2021-0108

82. Das A, Banik BK. 3 Semiconductor Characteristics of Tellurium and Its Implementations. In: *3 Semiconductor Characteristics of Tellurium and Its Implementations*. De Gruyter; 2022:55–84. doi:10.1515/9783110735840-003

83. Das A, Das A, Banik BK Tellurium-based chemical sensors. *Phys Sci Rev* 2022;8(12):4461–4501. Published online May 17, 2022. doi:10.1515/psr-2021-0116

84. Das A, Das A, Banik BK. 9 Tellurium-Based Chemical Sensors. In: *9 Tellurium-Based Chemical Sensors*. De Gruyter; 2022:183–224. doi:10.1515/9783110735840-009

85. Das A, Ray D, Banik BK. Tellurium in carbohydrate synthesis. *Phys Sci Rev* 2022;8(11):4157–4178. Published online May 7, 2022. doi:10.1515/psr-2021-0109

86. Das A, Ray D, Banik BK. 4 Tellurium in Carbohydrate Synthesis. In: *4 Tellurium in Carbohydrate Synthesis*. De Gruyter; 2022:85–106. doi:10.1515/9783110735840-004

87. Aldawood SAA, Das A, Banik BK. Tellurium-induced cyclization of olefinic compounds. *Phys Sci Rev*. 2022;8(12):4569–4609. Published online May 17, 2022. doi:10.1515/psr-2021-0119

88. Aldawood SAA, Das A, Banik BK. 11 Tellurium-Induced Cyclization of Olefinic *Compounds. In: 11 Tellurium-Induced Cyclization of Olefinic Compounds*. De Gruyter; 2022:249–290. doi:10.1515/9783110735840-011

89. Ray D, Das A, Mazumdar S, Banik BK. Tellurium-induced functional group activation. *Phys Sci Rev*. 2023;8(12):4821–4838. Published online June 2, 2022. doi:10.1515/psr-2021-0221

90. Ray D, Das A, Mazumdar S, Banik BK. 12 Tellurium-Induced Functional Group Activation. In: *12 Tellurium-Induced Functional Group Activation*. De Gruyter; 2022:291–308. doi:10.1515/9783110735840-012
91. Revol G, McCallum T, Morin M, Gagosz F, Barriault L. Photoredox Transformations with Dimeric Gold Complexes. *Angew Chem Int Ed.* 2013; 52(50):13342–13345. doi:10.1002/anie.201306727
92. Prier CK, Rankic DA, MacMillan DWC. Visible light photoredox catalysis with transition metal complexes: applications in organic synthesis. *Chem Rev.* 2013;113(7):5322–5363. doi:10.1021/cr300503r
93. Shi L, Xia W. Photoredox functionalization of C–H bonds adjacent to a nitrogen atom. *Chem Soc Rev.* 2012;41(23):7687–7697. doi:10.1039/C2CS35203F
94. Fukuzumi S, Ohkubo K. Organic synthetic transformations using organic dyes as photoredox catalysts. *Org Biomol Chem.* 2014;12(32):6059–6071. doi:10. 1039/C4OB00843J
95. Das A. LED Light sources in organic synthesis: An entry to a novel approach. *Lett Org Chem.* 2022;19(4):283–292.
96. Das A, Banik BK. Sustainable reactions in the synthesis of heterocycles. *Curr Organocatalysis.* 2022;9(1):3–3. doi:10.2174/2213337209012203281645523
97. Yadav RN, Shaikh AL, Das A, Ray D, Banik BK. Asymmetric synthesis of 3-pyrrole substituted β-lactams through p-toluene sulphonic acid-catalyzed reaction of azetidine-2,3-diones with hydroxyprolines. *Curr Organocatalysis.* 2022;9(4):337–345.
98. Das A, Yadav RN, Banik BK. Conceptual design and cost-efficient environmentally benign synthesis of beta-lactams. *Phys Sci Rev.* 2022;8(11):4053–4084. Published online May 4, 2022. doi:10.1515/psr-2021-0088
99. Das A, Yadav R, Banik BK. 10 Conceptual Design and Cost-Efficient Environmentally Benign Synthesis of Betalactams. In: *10 Conceptual Design and Cost-Efficient Environmentally Benign Synthesis of Betalactams.* De Gruyter; 2022:357–388. doi:10.1515/9783110797428-010
100. Miyazawa K, Koike T, Akita M. Hydroaminomethylation of olefins with aminomethyltrifluoroborate by photoredox catalysis. *Adv Synth Catal.* 2014;356(13):2749–2755. doi:10.1002/adsc.201400556
101. Tschudy D, Collins A. Malonic ester synthesis of δ-aminolevulinic acid. the reaction of n-3-bromoacetonylphthalimide with malonic ester. *J Org Chem.* 1959;24(4):556–557. doi:10.1021/jo01086a600
102. Shuto S, Takada H, Mochizuki D, Tsujita R, Hase Y, Ono S, et al. (.+-.)-(Z)-2-(Aminomethyl)-1-phenylcyclopropanecarboxamide derivatives as a new prototype of NMDA receptor antagonists. *J Med Chem.* 1995;38(15):2964–2968. doi:10.1021/jm00015a019
103. Bonnaud B, Cousse H, Mouzin G, Briley M, Stenger A, Fauran F, et al. 1-Aryl-2-(aminomethyl)cyclopropanecarboxylic acid derivatives. A new series of potential antidepressants. *J Med Chem.* 1987;30(2):318–325. doi:10.1021/jm00385a013
104. Yoon TP, Ischay MA, Du J. Visible light photocatalysis as a greener approach to photochemical synthesis. *Nat Chem.* 2010;2(7):527–532. doi:10.1038/nchem.687

105. McNally A, Prier CK, MacMillan DWC. Discovery of an α-amino C-H arylation reaction using the strategy of accelerated serendipity. *Science.* 2011;334(6059):1114–1117. doi:10.1126/science.1213920

106. Condie AG, González-Gómez JC, Stephenson CRJ. Visible-light photoredox catalysis: aza-Henry reactions via C-H functionalization. *J Am Chem Soc.* 2010;132(5):1464–1465. doi:10.1021/ja909145y

107. Zou YQ, Lu LQ, Fu L, Chang NJ, Rong J, Chen JR, et al. Visible-light-induced oxidation/[3+2] cycloaddition/oxidative aromatization sequence: a photocatalytic strategy to construct pyrrolo[2,1-a]isoquinolines. *Angew Chem.* 2011;123(31):7309–7313. doi:10.1002/ange.201102306

108. Koike T, Akita M. Photoinduced oxyamination of enamines and aldehydes with TEMPO catalyzed by [Ru(bpy)3]2+. *Chem Lett.* 2009;38(2):166–167. doi:10.1246/cl.2009.166

109. Allen LJ, Cabrera PJ, Lee M, Sanford MS. N-Acyloxyphthalimides as nitrogen radical precursors in the visible light photocatalyzed room temperature C-H amination of arenes and heteroarenes. *J Am Chem Soc.* 2014;136(15):5607–5610. doi:10.1021/ja501906x

110. Zuo Z, MacMillan DWC. Decarboxylative arylation of α-amino acids via photoredox catalysis: a one-step conversion of biomass to drug pharmacophore. *J Am Chem Soc.* 2014;136(14):5257–5260. doi:10.1021/ja501621q

111. Chinzei T, Miyazawa K, Yasu Y, Koike T, Akita M. Redox-economical radical generation from organoborates and carboxylic acids by organic photoredox catalysis. *RSC Adv.* 2015;5(27):21297–21300. doi:10.1039/C5RA01826A

112. Yamada K, Okada M, Fukuyama T, Ravelli D, Fagnoni M, Ryu I. Photocatalyzed site-selective C–H to C–C conversion of aliphatic nitriles. *Org Lett.* 2015;17(5):1292–1295. doi:10.1021/acs.orglett.5b00282

113. Ryu I, Tani A, Fukuyama T, Ravelli D, Montanaro S, Fagnoni M. Efficient C–H/C–N and C–H/C–CO–N conversion via decatungstate-photoinduced alkylation of diisopropyl azodicarboxylate. *Org Lett.* 2013;15(10):2554–2557. doi:10.1021/ol401061v

114. Okada M, Fukuyama T, Yamada K, Ryu I, Ravelli D, Fagnoni M. Sunlight photocatalyzed regioselective β-alkylation and acylation of cyclopentanones. *Chem Sci.* 2014;5(7):2893–2898. doi:10.1039/C4SC01072H

115. Arceo E, Jurberg ID, Álvarez-Fernández A, Melchiorre P. Photochemical activity of a key donor–acceptor complex can drive stereoselective catalytic α-alkylation of aldehydes. *Nature Chem.* 2013;5(9):750–756. doi:10.1038/nchem.1727

116. Martins A, Mariampillai B, Lautens M. Synthesis in the key of Catellani: norbornene-mediated ortho C-H functionalization. *Top Curr Chem.* 2010; 292:1–33. doi:10.1007/128_2009_13

117. Rouquet G, Chatani N. Catalytic functionalization of C(sp2)-H and C(sp3)-H bonds by using bidentate directing groups. *Angew Chem Int Ed Engl.* 2013;52(45):11726–11743. doi:10.1002/anie.201301451

118. Hoshikawa T, Inoue M. Photoinduced direct 4-pyridination of C(sp3)–H bonds. *Chem Sci.* 2013;4(8):3118–3123. doi:10.1039/C3SC51080H

119. Pirnot MT, Rankic DA, Martin DBC, MacMillan DWC. Photoredox activation for the direct β-arylation of ketones and aldehydes. *Science.* 2013;339(6127):1593–1596. doi:10.1126/science.1232993

120. Huang Z, Dong G. Catalytic direct β-arylation of simple ketones with aryl iodides. *J Am Chem Soc.* 2013;135(47):17747–17750. doi:10.1021/ja410389a

121. Shiraishi Y, Sakamoto H, Sugano Y, Ichikawa S, Hirai T. Pt–Cu bimetallic alloy nanoparticles supported on anatase TiO2: Highly active catalysts for aerobic oxidation driven by visible light. *ACS Nano.* 2013;7(10):9287–9297. doi:10.1021/nn403954p

122. Lian M, Li Z, Cai Y, Meng Q, Gao Z. Enantioselective photooxygenation of β-keto esters by chiral phase-transfer catalysis using molecular oxygen. *Chem Asian J.* 2012;7(9):2019–2023. doi:10.1002/asia.201200358

123. Higashimoto S, Kitao N, Yoshida N, Sakura T, Azuma M, Ohue H, et al. Selective photocatalytic oxidation of benzyl alcohol and its derivatives into corresponding aldehydes by molecular oxygen on titanium dioxide under visible light irradiation. *J Catal.* 2009;266(2):279–285. doi:10.1016/j.jcat.2009.06.018

124. Zhang Q, Jiang L, Wang J, Zhu Y, Pu Y, Dai W. Photocatalytic degradation of tetracycline antibiotics using three-dimensional network structure perylene diimide supramolecular organic photocatalyst under visible-light irradiation. *Appl Catal B: Environ.* 2020;277:119122. doi:10.1016/j.apcatb.2020.119122

125. Liu L, Huang J, Yu H, Wan J, Liu L, Yi K, et al. Construction of MoO3 nanopaticles/g-C3N4 nanosheets 0D/2D heterojuntion photocatalysts for enhanced photocatalytic degradation of antibiotic pollutant. *Chemosphere.* 2021;282:131049. doi:10.1016/j.chemosphere.2021.131049

126. Jin D, He D, Lv Y, Zhang K, Zhang Z, Yang H, et al. Preparation of metal-free BP/CN photocatalyst with enhanced ability for photocatalytic tetracycline degradation. *Chemosphere.* 2022;290:133317. doi:10.1016/j.chemosphere.2021.133317

127. Doosti M, Jahanshahi R, Laleh S, Sobhani S, Sansano JM. Solar light induced photocatalytic degradation of tetracycline in the presence of ZnO/NiFe$_2$O$_4$/Co$_3$O$_4$ as a new and highly efficient magnetically separable photocatalyst. *Frontiers in Chemistry.* 2022;10. https://www.frontiersin.org/articles/10.3389/fchem.2022.1013349. Accessed February 13, 2024.

128. Fatiadi AJ. Preparation and Synthetic Applications of Cyano Compounds. In: *Triple-Bonded Functional Groups (1983).* John Wiley & Sons, Ltd; 1983:1057–1303. doi:10.1002/9780470771709.ch9

129. Fleming FF, Yao L, Ravikumar PC, Funk L, Shook BC. Nitrile-containing pharmaceuticals: efficacious roles of the nitrile pharmacophore. *J Med Chem.* 2010;53(22):7902–7917. doi:10.1021/jm100762r

130. Miller JS, Manson JL. Designer magnets containing cyanides and nitriles. *Acc Chem Res.* 2001;34(7):563–570. doi:10.1021/ar0000354

131. Sandmeyer T. Ueberführung der drei Nitraniline in die Nitrobenzoësäuren. *Ber Dtsch Chem Ges.* 1885;18(1):1492–1496. doi:10.1002/cber.188501801322

132. Zhou S, Junge K, Addis D, Das S, Beller M. A general and convenient catalytic synthesis of nitriles from amides and silanes. *Org Lett.* 2009;11(11):2461–2464. doi:10.1021/ol900716q

133. Yamaguchi K, Fujiwara H, Ogasawara Y, Kotani M, Mizuno N. A tungsten–tin mixed hydroxide as an efficient heterogeneous catalyst for dehydration of aldoximes to nitriles. *Angew Chem Int Ed.* 2007;46(21):3922–3925. doi:10.1002/anie.200605004

134. Oishi T, Yamaguchi K, Mizuno N. Catalytic oxidative synthesis of nitriles directly from primary alcohols and ammonia. *Angew Chem Int Ed.* 2009;48(34):6286–6288. doi:10.1002/anie.200900418

135. Yamaguchi K, Mizuno N. Scope, kinetics, and mechanistic aspects of aerobic oxidations catalyzed by ruthenium supported on alumina. *Chem Eur J.* 2003;9(18):4353–4361. doi:10.1002/chem.200304916

136. Uddin MT, Nicolas Y, Olivier C, Toupance T, Müller MM, Kleebe HJ, et al. Preparation of RuO$_2$/TiO$_2$ mesoporous heterostructures and rationalization of their enhanced photocatalytic properties by band alignment investigations. *J Phys Chem C.* 2013;117(42):22098–22110. doi:10.1021/jp407539c

137. Bang S, Lee S, Park T, Ko Y, Shin S, Yim SY, et al. Dual optical functionality of local surface plasmon resonance for RuO2 nanoparticle–ZnO nanorod hybrids grown by atomic layer deposition. *J Mater Chem.* 2012;22(28):14141–14148. doi:10.1039/C2JM31513K

138. Ovoshchnikov DS, Donoeva BG, Golovko VB. Visible-light-driven aerobic oxidation of amines to nitriles over hydrous ruthenium oxide supported on TiO2. *ACS Catal.* 2015;5(1):34–38. doi:10.1021/cs501186n

Novel LED-Induced Photochemical Reactions Toward the Synthesis of Organic Compounds

3

Aparna Das and Bimal Krishna Banik

3.1 INTRODUCTION

An important role is played by photochemical reactions in the environment. In photocatalysis, light is combined with a catalyst that absorbs light for a chemical reaction to occur, resulting in a change in the rate of chemical transformation. This green chemistry pathway has many possible applications, including antibacterial purposes, wastewater treatment, deodorizing effects, air purification, antifogging effects, the ability to self-clean, water purification, and the ability to remediate water. Photocatalytic technology offers a combination of

DOI: 10.1201/9781003634249-3

environmental benefits, low cost, and sustainability. As a continuation of our research into mechanistically crucial syntheses using microwave-irradiation in accordance with green chemistry principles, we have done a comprehensive review of the available photochemical methods.

There is also growing interest in photochemical technology due to its exceptional capability to degrade toxic compounds during solar irradiation, which is greatly reducing environmental problems as a result of its use. The rate at which photochemical reactions occur varies with the molecular structure of the compounds, how strong the light source is, and whether or not other reactants are present.

This chapter aims to provide a brief overview of the organic reactions that have been catalyzed by different types of light-emitting diode (LED) light sources [ultraviolet (UV) LEDs, blue LEDs, and white LEDs], as well as photocatalysts (transition metal complexes and organic dyes). A detailed discussion is given on the effects of white, UV, and blue LEDs in reactions such as reduction, oxidation, cycloaddition, sensitization, and isomerization. The study examines the yield and rate of reaction produced by LED lights as well as other sources of light such as household bulbs, CFLs, and halogen lamps. A review of the most recent important studies will be presented in this chapter. This field has also been discussed in terms of prospects and challenges. In addition to photoinduced electron transfers, energy transfer processes are discussed in the paper[1,2]. As far as we know, no other chapter have been published on the photocatalysis of organic compounds using UV, white, and blue LED sources. Based on all the facts, this chapter is extremely important and timely.

3.2 NATURE OF THE ELECTROMAGNETIC RADIATION

An electromagnetic (EM) spectrum is the entire range of EM radiation (**Figure 3.1**). It consists of radio waves, microwaves, infrared (IR), visible light, UV, X-rays, and gamma rays, which have wavelengths ranging from 100 km to 10^{-9} nm. Solar radiation is the energy emitted by the Sun that travels through space in all directions as EM waves. Climate and atmospheric processes in earth are influenced by this energy. It can be seen as UV rays, visible light waves, IR waves, and other types of energy. As this incoming radiation enters the atmosphere, some of it is reflected off the clouds, some is absorbed and some passes through to the surface of the Earth.

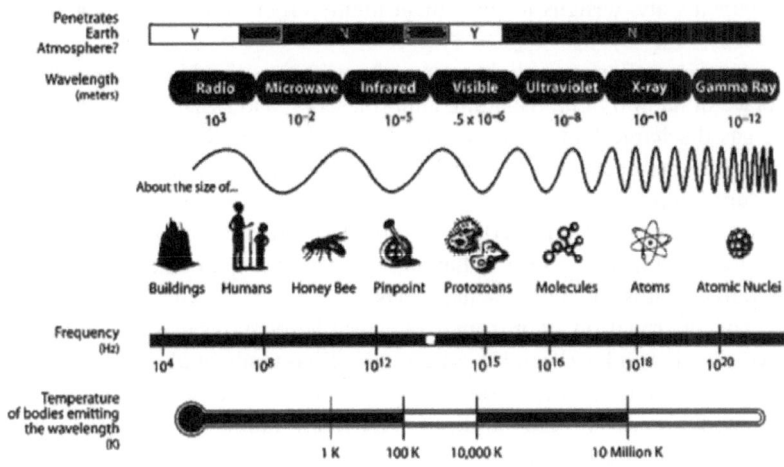

FIGURE 3.1 Schematic of the EM spectrum that depicts various properties. (Adapted from public domain.)

3.3 PRELIMINARY FACTS ON PHOTOCHEMICAL REACTIONS

Organic compounds can be structurally altered by EM radiation in the UV and visible ranges. The energy of visible light is lower than that of UV light. Organic molecules with molecularly complex structures are more resistant to photodecomposition in low-energy visible light. It has been possible to synthesize a wide range of useful and important carbocycles and heterocycles through visible-light-induced photocatalysis[3–14]. With photocatalytic reactions, organic compounds can generate molecular skeletons that are difficult to obtain via ground-state reactions[15–17]. Transition metal catalysts that are visible-light-activated have attracted considerable research attention[18–25]. Different sources of irradiation were used for photocatalysis. In addition to natural light, artificial light is also included.

Using natural light sources for photocatalysis involves activating a photocatalyst with sufficiently energetic photons to generate electron-hole pairs, which then produce the intermediate. In the literature, this area of photocatalysis is called solar photocatalysis. Solar radiation is an advantageous source of light in photocatalysis to avoid high costs of artificial illumination (lamps and electricity). It is, however, cost-intensive and requires complex components and designs to implement highly efficient solar photocatalysis systems.

A variety of artificial light sources have been utilized in photocatalysis in place of natural light, including lamps, lasers[26], and LEDs[27]. Incandescent and gas discharge lamps are the two types of conventional lamps used in photocatalytic reactions. There are three types of gas discharge lamps: low-pressure lamps, such as fluorescent lamps, high-pressure lamps, and high-intensity discharge lamps (HID). The light spectrum emitted by lamps is narrower than that of solar irradiation; however, it is quite wide when compared to the emission of lasers and LEDs.

The degradation of nitrobenzene by photocatalysis was compared between solar and artificial irradiation by Bhaktende et al.[28]. Under artificial UV radiation, the photocatalytic reaction rate was much higher than under concentrated sunlight. Artificial lamps contain a higher percentage of useful components than sunlight, which increases their usefulness. Solar irradiation, on the other hand, can provide higher efficiency for photocatalytic processes when the catalyst uses both visible light and UV light[29].

A laser produces coherent, monochromatic, high-intensity light with low beam divergence, and its wavelength can be altered to the desired range. Despite their advantages, lasers have not been widely used in photocatalysis because of their high cost, specialized training requirement, unsuitability for reactor design, and safety concerns. However, laser irradiated photocatalysis is also gaining general interest since laser beams produce unique light properties that have previously been shown to enhance photocatalytic efficiency[30].

3.4 WORKING PRINCIPLE OF LED

To produce light, LEDs use inorganic semiconductors in the form of junctions with electron-carrying n-layers and hole-carrying p-layers which are normally stacked to create solid-state light sources[31]. When a forward voltage is applied to the n-layer and the p-layer, electrons are ejected from the respective layers, and the electrons and holes that are ejected are recombined within the device to release photons (**Figure 3.2**). Semiconductor materials have different bandgap energies, which determine the color of the light they produce. There is a strong correlation between light intensity and forward voltage, whereas photon energy is a function of wavelength.

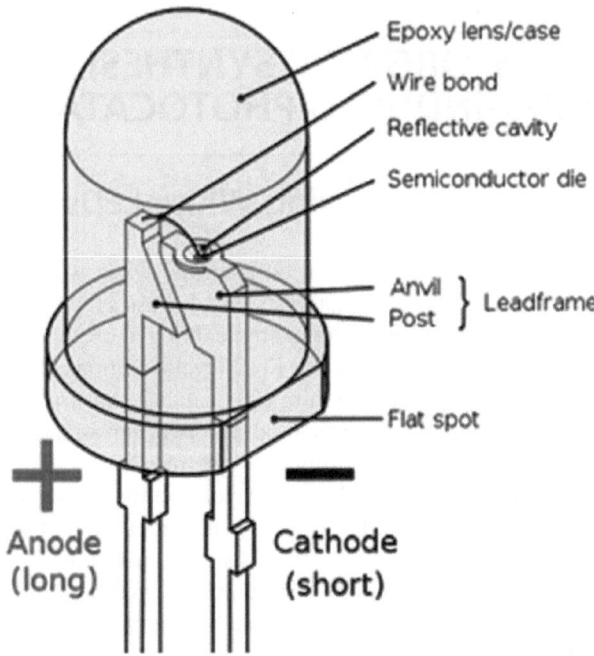

FIGURE 3.2 A labelled drawing of the normal type LED. (Adapted from public domain.)

LED light intensity increases as the forward voltage is increased, while the energy of the generated photons increases as the wavelength of the LED light decreases.

LEDs are a superior light source in many ways compared to conventional lamps. They are more affordable, durable, compact, lightweight, and operate at a lower temperature. Due to the losses of photon energy during the conversion process as well as the losses of heat generated by the filament, fluorescent and incandescent lamps are not as efficient as they could be. Solid-state lighting (SSL) sources such as LEDs, however, become exponentially more efficient as semiconductor technologies, materials science, and optics advance. Therefore, LEDs are a good choice of light source for photocatalytic applications. The advantages of LEDs over other light sources include their efficiency, power, compatibility, lifetime, and environmental friendliness. LEDs provide new possibilities for photochemical reactions while consuming less energy than conventional lamps because of their substantial properties.

3.5 ORGANIC SYNTHESIS BY LEDS-INDUCED PHOTOCATALYSIS

3.5.1 Photocatalysis using White LEDs

Several photochemical reactions can be achieved when white LEDs are combined with appropriate photocatalysts. We discuss a few selected reactions here. By using white LEDs, for example, bridged cyclobutanes could be synthesized through intramolecular [2+2] cycloadditions of dienones[32]. With the help of a polypyridyl iridium (III) catalyst, the $Ir(dF(CF_3)ppy)_2$ (dtbbpy)PF$_6$, the reaction was promoted through sensitization. In this experiment, an effective cross addition was obtained rather than a straight addition under light irradiation. Benzobicycloheptanone-bridged products were produced in high yields, up to 96%, with excellent regioselectivity. Schematic representation of the reaction is shown in **Figure 3.3**. Various spectroscopic, electrochemical, and computational methods confirmed the existence of triplet energy transfer in visible light indices.

Our research group has been working on diverse medicinally active compounds. These have culminated in the synthesis of novel sugars, natural products, different types of carbocyclic and heterocyclic organic compounds[33-40].

In many natural products and biologically active compounds, indoles and their derivatives are commonly found[41-45]. There is great importance to 2-substituted indoles among them due to their biological activities[46-50].

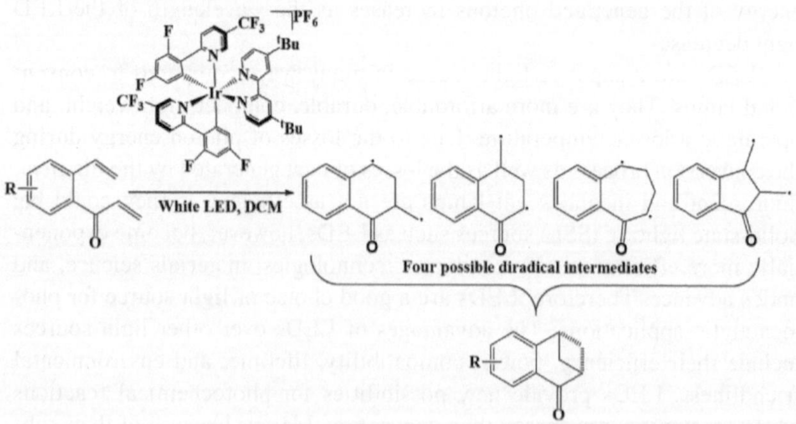

FIGURE 3.3   Intramolecular crossed [2+2] photocycloaddition reaction.

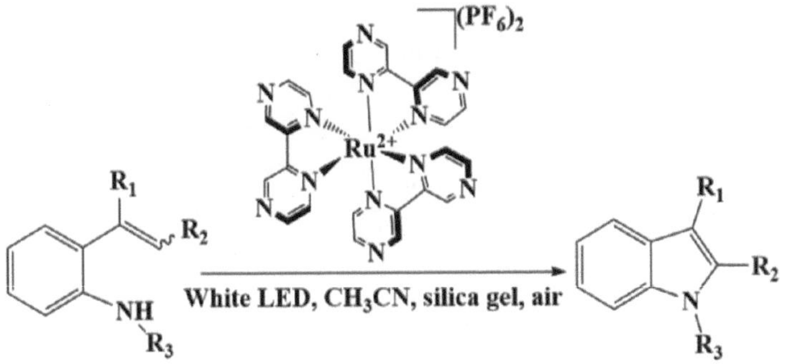

FIGURE 3.4   Preparation of N-arylindoles.

A number of researchers have also reported that indole can be synthesized via visible-light-induced photocatalysis. Maity et al., for instance, synthesized indole by oxidatively creating C-N bonds and aromaticizing styrylanilines under visible light[51]. The synthesis was carried out using white LEDs and $[Ru(bpz)_3](PF_6)_2$ photocatalyst as shown in **Figure 3.4**.

Our endeavor on various catalytic procedures has become highly significant. During this study, we have also seen the powerful effects of photo radiation in synthesis[52–58].

As a photocatalytic method for synthesis of 2-substituted N-free indoles using white LEDs, intermolecular cyclization of styryl azides was reported with good to excellent results[59]. The reaction was carried out using ruthenium complex (Ru(bpy)3Cl2·6H2O) as a photocatalyst at room temperature (**Figure 3.5**). Additionally, the possible reaction routes leading to the

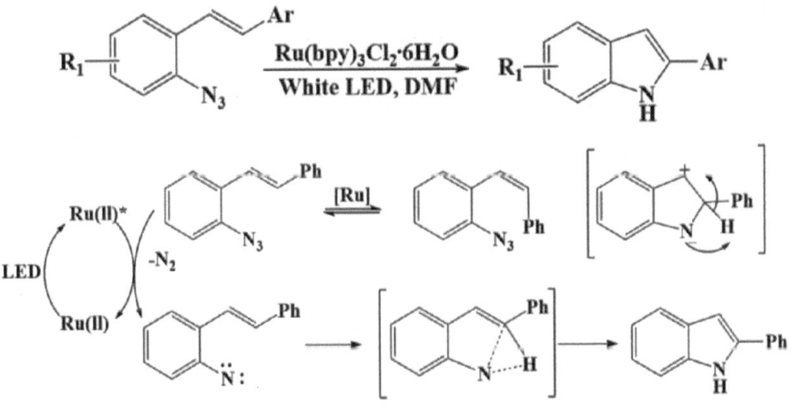

FIGURE 3.5   Styryl azide's sensitization.

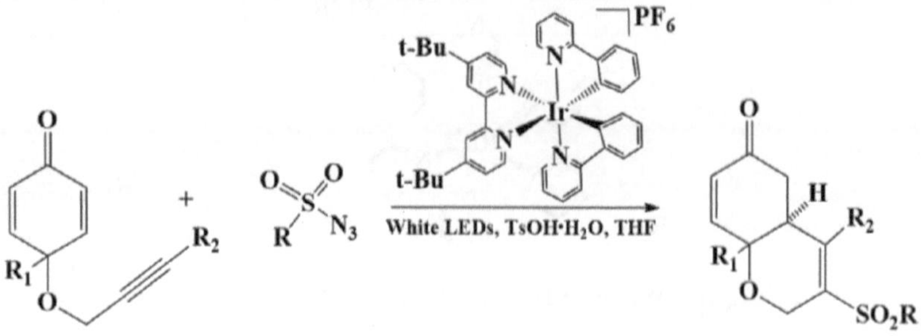

FIGURE 3.6   Sulfonyl azide's sensitization.

formation of indoles were examined. Based on the cyclic voltammetry data, it appears that the photocatalyst failed to give an electron for the reduction of the substrate. Thus, an energy transfer may be a possible mechanism for this transformation.

The intermolecular cyclization of azides has also been reported to use visible light photocatalysis. As reported by Zhu et al., sulfonylative and azidosulfonylative cyclization of sulfonyl azides was achieved by visible-light-photosensitization[60] (**Figure 3.6**). A white LED and Ir(ppy)$_2$(dtbbpy)PF$_6$ photocatalyst were used to trigger the reaction. Radical initiation is described as the formation of triplet sulfonyl nitrene upon photosensitization and the generation of tetrahydrofuran-2-yl radicals by hydrogen atoms from THF. The sulfonyl radical is formed by the interaction of the radical with the sulfonyl azide.

Photoredox functionalization of 3,4-dihydro-1,4-benzoxazin-2-ones using indoles was demonstrated (**Figure 3.7**)[61]. During the reaction, 9,10-phenanthrenedione and a Lewis acid, Zn(OTf)$_2$, were both used as catalysts. A similar reaction can be performed in other arenes that are rich in electrons. It was revealed that the mechanism proposed involved single-electron transfer.

It has been shown that N-bromosuccinimide, catalyzed by erythrosine B (organic dye), is capable of photoredox bromination in good to excellent yields of an array of arenes and heteroarenes (**Figure 3.8**)[62]. The bromination reaction was investigated under a variety of light sources. A moderate

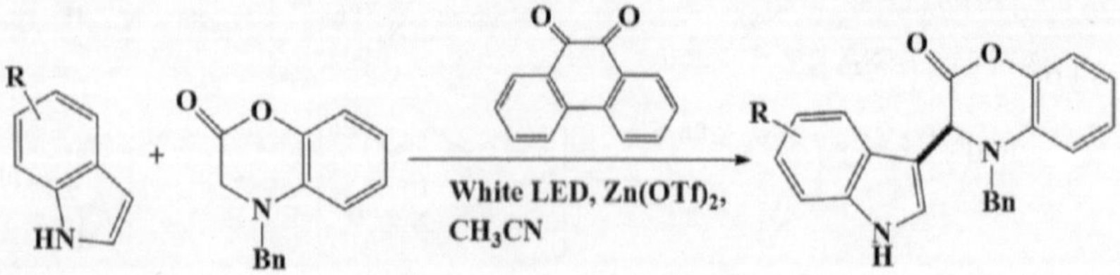

FIGURE 3.7   Friedel–Crafts reaction with various indoles.

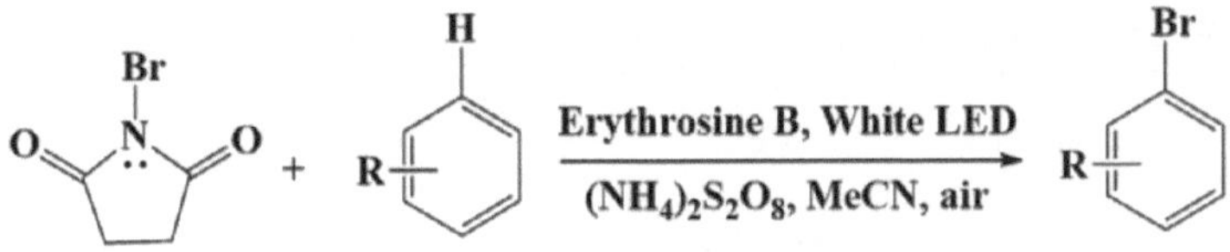

FIGURE 3.8 Bromination utilizing N-bromosuccinimide.

reduction in yield was observed when blue LEDs were used. An LED light chamber with green or red LEDs, however, had a severe adverse impact on the bromination reaction. It was found that white LEDs provided the highest yields, and they were considered to be the most reliable. Based on the mechanistic studies, it appears that N-bromosuccinimide is oxidized by erythrosine B by a single electron.

It has been reported that hydrocarbons can be bromined with CBr_4 by irradiating them with light emitting diodes[63]. CBr_4 is used as a bromine source for the bromination of hydrocarbons which is induced by white LED irradiation (**Tables 3.1**, **3.2 and Figure 3.9**). Using no catalysts, additives,

TABLE 3.1 Bromination of cyclohexane employing CBr4 under irradiation

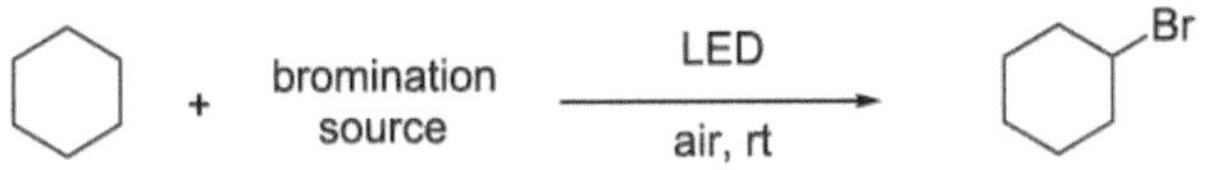

ENTRY	BROMINATION SOURCE (mmol)	TIME (h)	YIELD (%)
1	CBr$_4$ (0.20)	2	77
2	CBr$_4$ (0.20)	3	108
3	CBr$_4$ (0.20)	4	148
4	CBr$_4$ (0.20)	5	148
5	CBr$_4$ (0.20)	24	150
6	CHBr$_3$ (0.20)	24	27
7	CH$_2$Br$_2$ (0.20)	24	0
8	NBS (0.20)	24	31
9	Bu$_4$NBr (0.20)	24	0
10	CBr$_4$ (0.02)/CHBr$_3$ (0.20)	24	39
11	CBr$_4$ (0.02)/CH$_2$Br$_2$ (0.20)	24	16
12	CBr$_4$ (0.02)/NBS (0.20)	24	79
13[c]	CBr$_4$ (0.20)	24	0
14[d]	CBr$_4$ (0.20)	24	87

Source: Adapted with permission from ref. [63].

TABLE 3.2   Bromination of other substrates utilizing CBr_4 under irradiation

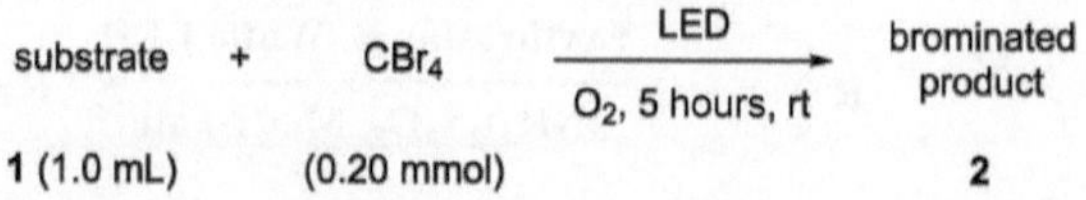

substrate	+	CBr_4	$\xrightarrow[O_2,\ 5\ hours,\ rt]{LED}$	brominated product
1 (1.0 mL)		(0.20 mmol)		**2**

ENTRY	SUBSTRATE	PRODUCT	YIELD (%)
1	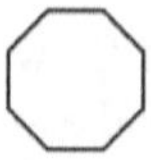	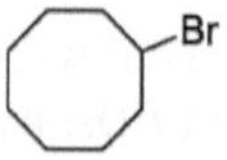 Br	178
2		 Br	14
		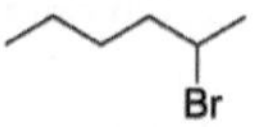 Br	84
		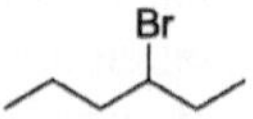 Br	41
3	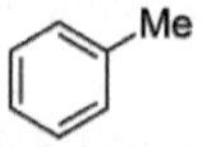 Me	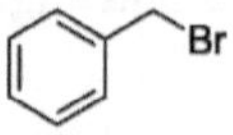 Br	140

Adapted with permission from ref. [63].

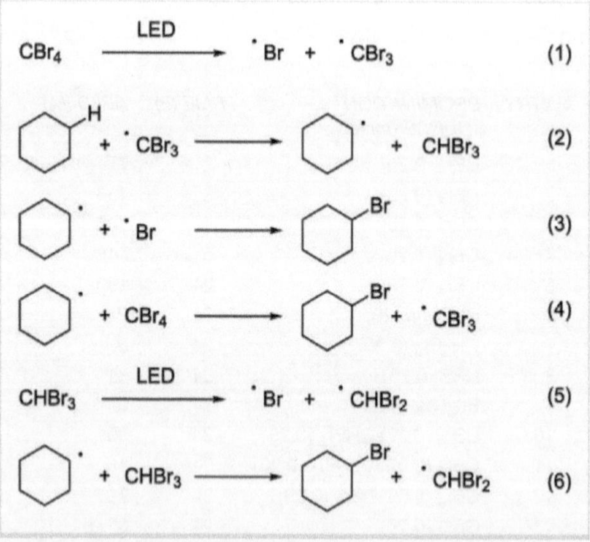

FIGURE 3.9   Plausible mechanism for bromination. (Adapted with permission from ref. [63].)

inert conditions, or heating, monobromides were synthesized efficiently. CBr4 absorbs light according to its action and absorption spectra, thereby producing active species for bromination. The generation of $CHBr_3$ has been confirmed by Nuclear Magnetic Resonance (NMR) spectroscopy. According to Gas Chromatography Mass Spectrometry (GC-MS) spectrometry analysis, the bromination reaction involves homolytic cleavage of the C-Br bonds in the CBr_4 compound, followed by radical abstraction of hydrogen atoms from the hydrocarbon molecules.

3.5.2 Photocatalysis using Blue LEDs

The following section discusses selected photochemical reactions using blue LEDs as light sources. The cyclobutabenzocypyranones are essential building blocks for tetrahydro dibenzofurans that are biologically active. Cycloaddition is performed by photocatalysts. Our group has prepared the synthesis numerous beta lactams by cycloaddition reaction under thermal, photochemical and microwave-induced irradiation methods[64-75]. Microwave-induced energy for the preparation of diverse organic structures is used by our group[76-94]. Coumarin-3-carboxylates were cycloadditioned with acrylamide analogs at room temperature using a visible light protocol (**Figure 3.10**)[95]. In this experiment, a blue LED was used as the light source, and an iridium complex (FlrPic) was used as the photosensitizer for the formation of an array of cyclobutabenzocypyranones.

Ir(ppy)3 photocatalysts were used to carry out a highly regioselective, diastereoselective [2+2] cycloaddition between olefins and 1,4-dihydropyridines using visible light as a catalyst (**Figure 3.11**)[96]. An effective procedure in the excellent yield synthesize of polysubstituted 2-azabicyclo-[4.2.0] octanes is provided by this strategy. Based on the mechanistic analysis, an energy transfer pathway appears to be involved in the reaction.

FIGURE 3.10   Intermolecular [2+2] photocycloaddition reactions of acrylamides and coumarin-3-carboxylates.

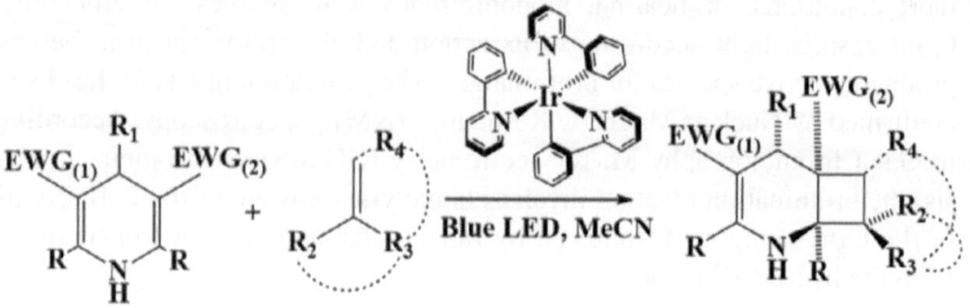

FIGURE 3.11   Cycloaddition between olefins and 1,4-dihydropyridines.

An asymmetric photoredox catalysis was also demonstrated using blue LEDs and chiral rhodium complexes[97]. The chiral Lewis acid in this protocol is directly activated by visible light to catalyze stereoselective [2+2] cycloaddition reactions (**Figure 3.12**). As a result, it served as both an antenna for harvesting visible light and as a chiral entity in the synthesis of complex

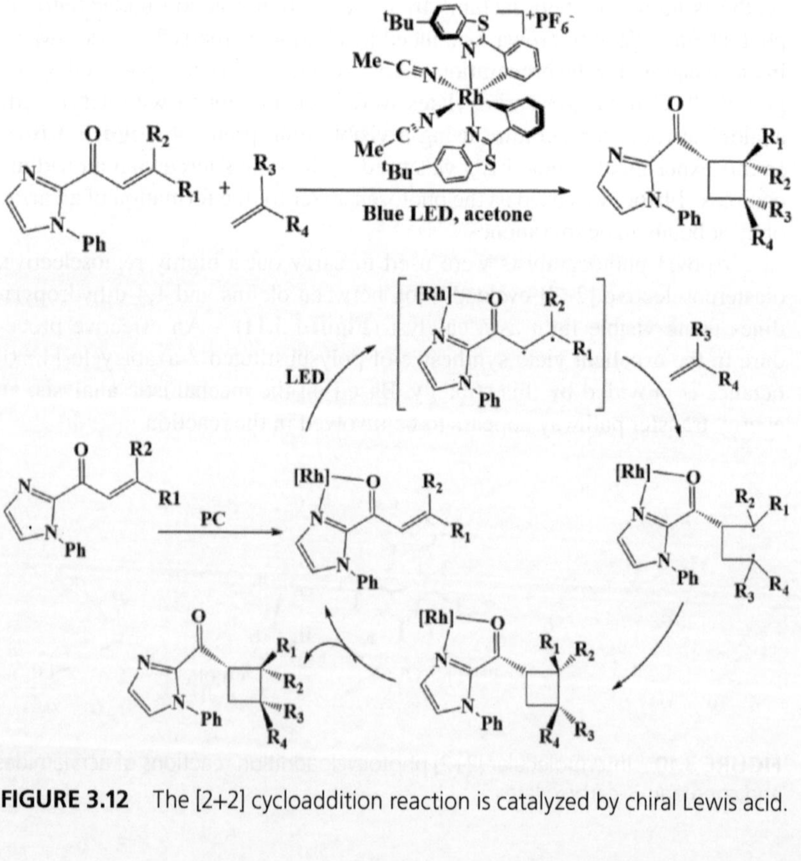

FIGURE 3.12   The [2+2] cycloaddition reaction is catalyzed by chiral Lewis acid.

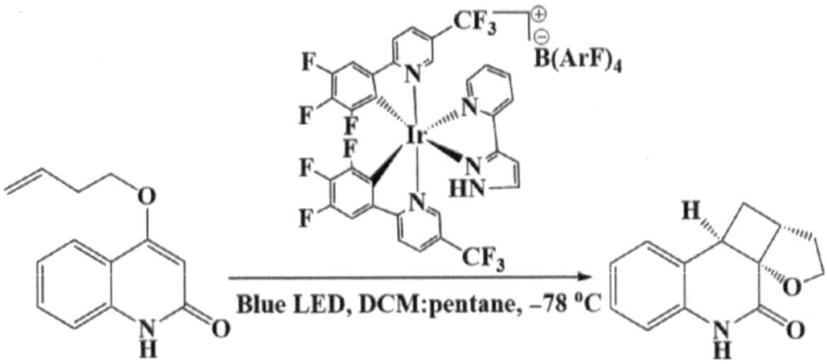

FIGURE 3.13   Chiral iridium catalyzed intramolecular [2+2] cycloaddition.

cyclobutanes that have high diastereo- and enantioselectivity. A mechanism for energy transfer has been demonstrated through mechanistic studies. Due to the fact that UV light was used to activate the catalysis of chiral Lewis acid, chiral oxazaborolidine promoted an enantioselective [2+2] photocycloaddition by redshifting the absorption of the substrate's UV light. There are a few limitations to this UV-light activation method, including its use of low temperatures, low UV-light intensity, and high catalyst loading.

It was reported that enantiopure iridium complexes were used as triplet sensitizers in an enantioselective intramolecular [2+2] photocycloaddition of coumarin involving hydrogen bonding interactions (**Figure 3.13**)[98]. It is efficient to transfer Dexter energy and induce stereoinduction in transition-metal photosensitizers by administering substrates within chiral environments.

During the course of our exploratory research, we became interested in using special properties of metals for diverse purposes. One of these metals is tellurium[99–110]. In the presence of organic dyes, [3+2] cycloaddition of 2H-azirines with alkynes could be accomplished under visible light irradiation (**Figure 3.14**)[111]. Pyrroles can be synthesized in good yields using this metal-free photocatalytic method. Pyrroles has been used to synthesize drug analogues. A significant effect of the light source was observed on the efficiency of the reaction. In an experiment using blue LEDs instead of white LEDs, yields increased dramatically and reaction times were shortened. As demonstrated in the control experiment, there was no reaction without either the photoredox catalyst or the light source.

Based on the mechanistic study, the initial step involves the energy transfer process that sensitizes vinyl azide to produce a triplet intermediate. After releasing nitrogen gas, nitrene forms as the main intermediate, which is then easily converted into azirine (**Figure 3.14**). In the last step, photoredox cycloaddition of the azirine intermediate take place in order to create the

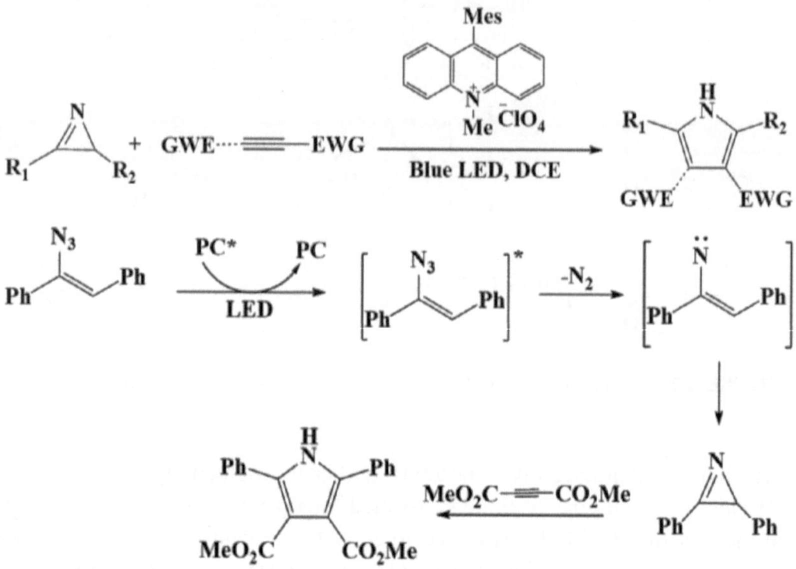

FIGURE 3.14   Photocatalytic [3+2] cycloaddition for the preparation of pyrroles.

corresponding product. Therefore, this reaction is the first instance of a photocascade reaction that combines single-electron transfer and energy transfer pathways.

By using $Ru(dtbbpy)_3(PF_6)_2$ as a photocatalyst and sensitizing aryl and vinyl azides with visible light, nitrenes were produced (**Figure 3.15**). Various C-N bond-forming reactions can be carried out on reactive nitrenes. Using this method, highly substituted pyrroles and indoles can be synthesized conveniently and efficiently[112]. The control experiments show that light and an appropriate photocatalyst are both necessary.

While the variety of visible-light sources can be used to perform the reactions, the reactions using blue LEDs were faster and yielded more than household light bulbs. When UV light was used for the same reaction, significant photodecomposition took place resulting in a poor mass balance and lower yields. In these types of reactions, visible light can indeed provide a significant advantage over other types of light. It has been demonstrated that the progression of the reaction occurs via a pathway of energy transfer.

An atom-economic and direct method of C-H amidation of heteroarenes using benzoyl azides requires blue LED (light source), $Ru(bpy)_3Cl_2 \cdot 6H_2O$ (a photocatalyst), and phosphoric acid (**Figure 3.16**)[113]. Under very mild reaction conditions, a good yield of hetero-aromatic amides was obtained with

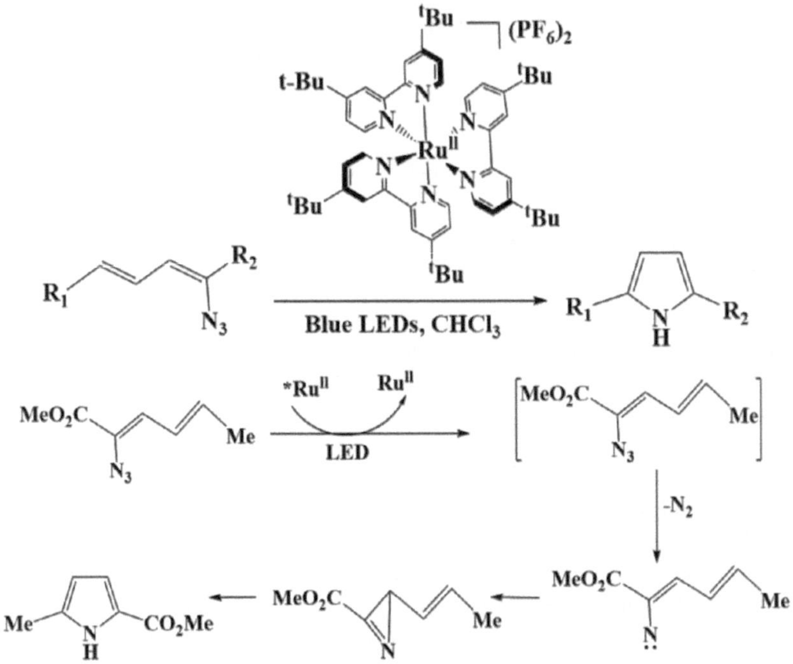

FIGURE 3.15 Preparation of pyrroles and indoles by triplet sensitization of azides.

dinitrogen the only by-product. It can be carried out with heteroaryl-, aryl-, or alkenyl acyl azides, and can be applied to heteroarenes such as indoles, pyrroles, furans, thiophenes, and benzofurans. In addition, oxazolines, dioxazoles, and aziridines were produced under these conditions when benzoyl azides were exposed to UV light. Under these conditions, no formal amidation or heterocycle conversion was observed.

FIGURE 3.16 C-H amidation of heteroarenes.

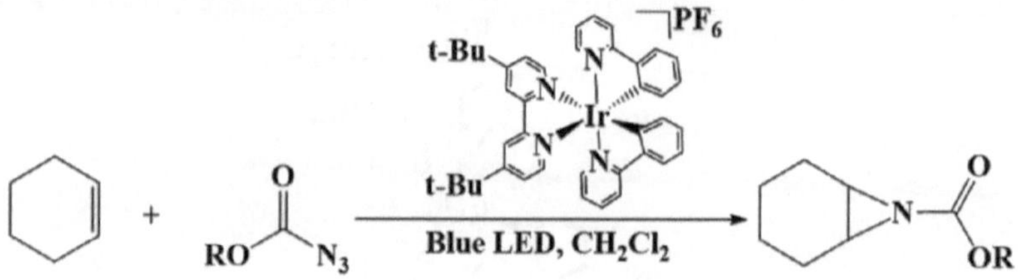

FIGURE 3.17   Intermolecular aziridination of alkenes.

As a result of the direct photochemical activation of azidoformates, aziridination and allylic amination products can be formed competitively. With visible-light-activated transition-metal complexes as the triplet sensitizer, aziridines can be selectively synthesized from azidoformates via spin-selective photogeneration of triplet nitrenes.[114]. Photocatalyst used to accomplish this was $Ir(ppy)_2(dtbbpy)PF_6$ complex and the light source was blue LED (**Figure 3.17**).

It has been reported that a DPZ (dicyanopyrazine-derived chromophore) can be used as a photocatalyst to initiate oxygenation reactions in indoles with visible light[115]. By changing the pH of the reaction or considering inorganic salts, isatins, or formylformanilides could be produced with satisfactory yields (**Figure 3.18**). Furthermore, benzoxazinones, formylformanilide derivatives, and 2,2-disubstituted 3-oxindoles were synthesized using this method.

Copper catalyzed visible light induced Ullmann-type C-N coupling reaction between aryl iodides and carbazole derivatives was reported[116] (**Figure 3.19**). This reaction took place under mild conditions with the use of $Ir(ppy)_3$, an iridium-based photocatalyst.

In the catalytic cycle, the key step was a reaction between the copper-carbazolide complex and the excited photocatalyst resulting in an EnT reaction. During this process, an excited copper-carbazolide complex is formed. A single electron transfer is then performed between the complex and aryl iodide in order to produce the CuII complex and the aryl radical. In the subsequent step,

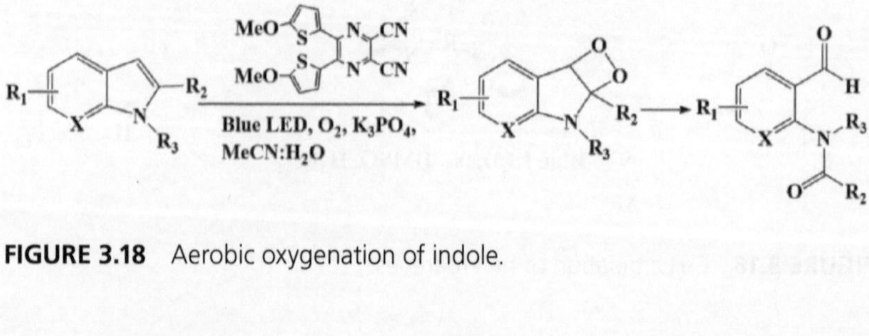

FIGURE 3.18   Aerobic oxygenation of indole.

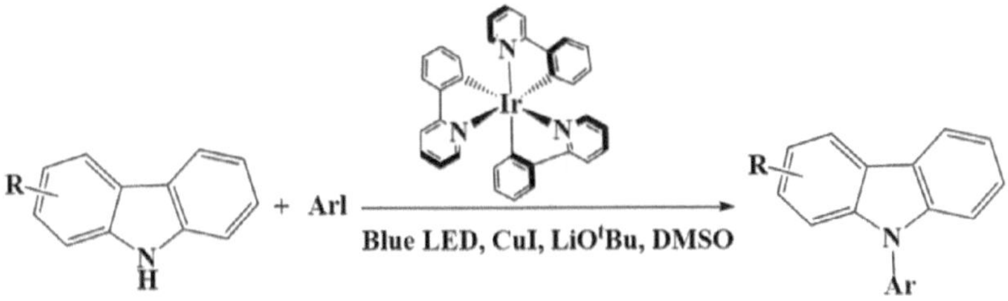

FIGURE 3.19 C-N coupling reaction between aryl iodides and carbazole derivatives.

the reaction between aryl radical and the intermediate took place to generate the C-N coupling product and regenerates the original CuI complex.

For fused heterocyclic systems synthesis, 6π cyclization induced by light is a crucial procedure. In the presence of an iridium complex, across a wide variety of substrates using the blue LED light promoted cyclic 2-aryloxyketones, an efficient and high yielding arylation can be achieved via an energy transfer pathway (**Figure 3.20**)[117]. Under these conditions, the 2-arylthioketones and the 2-arylaminoketones were also cyclized effectively.

A photoredox C-H functionalization reaction of tertiary amines, utilizing visible light and iridium complex as a catalyst, was reported for both intermolecular and intramolecular functionalization (**Figure 3.21**)[118]. With and without oxygen, the same starting material produced two different types of products. Despite the absence of oxygen, intermolecular additions of N, N-dimethylanilines to electron-deficient alkenes produced γ-amino nitriles in good to high yields. Under mild reaction conditions, tetrahydroquinoline derivatives were produced in good yields by a radical addition/cyclization reaction. Indole-3-carboxaldehyde derivatives were produced through the intramolecular version of the radical addition. A new photoredox catalyzed cleavage reaction of C-C bonds was discovered by mechanistic analysis of this reaction.

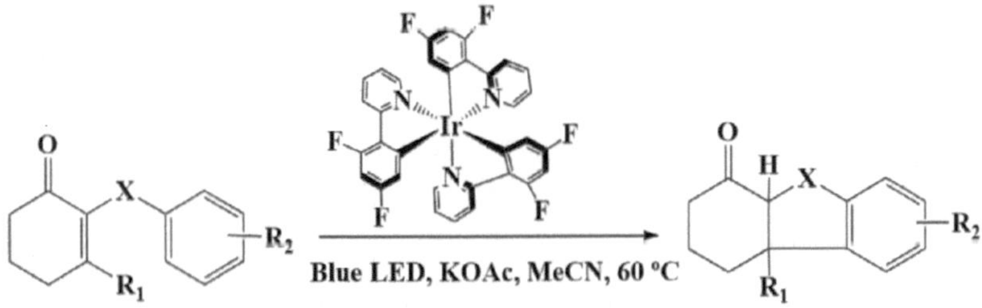

FIGURE 3.20 6π cyclization reaction.

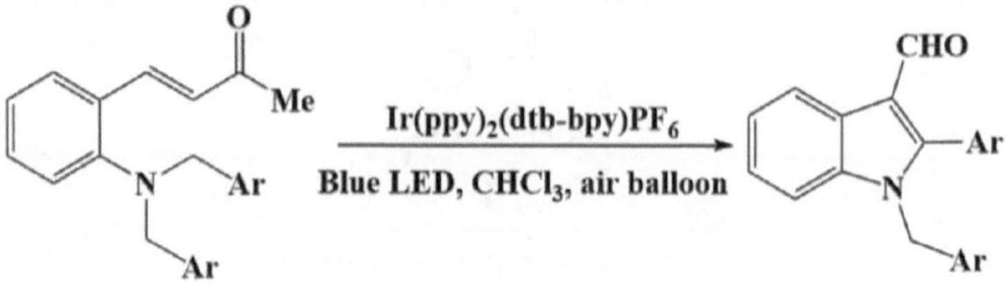

FIGURE 3.21   Preparation of N-alkylindoles.

By the utilization of photoredox catalysis, the hydroaminomethylation of olefins can be achieved with aminomethyltrifluoroborate[119]. Utilizing N-protected aminomethyltrifluoroborates, the development of a photocatalytic hydroaminomethylation of olefins has been demonstrated in this work. By using the proposed methodology, it is possible to introduce a primary aminomethyl group onto a C=C bond which was electron-deficient. By utilizing this reaction, it can be a simple means of obtaining many useful derivatives of aminobutyric acid (GABA) that can be synthesized, such as baclofen.

As a first step, in the presence of the Ir photoredox catalyst (1a), the photocatalytic reaction between Boc-protected aminomethyltrifluoroborate and methyl acrylate under blue LED's illumination for one hour under various solvent conditions was examined. The corresponding aminomethylated product 4aa was obtained as a result of the procedure in 92% NMR yield (**Figure 3.22**). A study

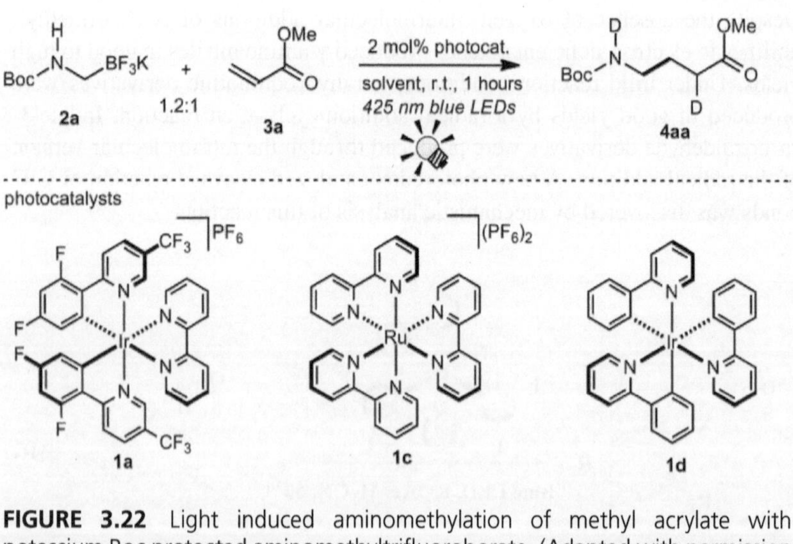

FIGURE 3.22   Light induced aminomethylation of methyl acrylate with potassium Boc-protected aminomethyltrifluoroborate. (Adapted with permission from ref. [119].)

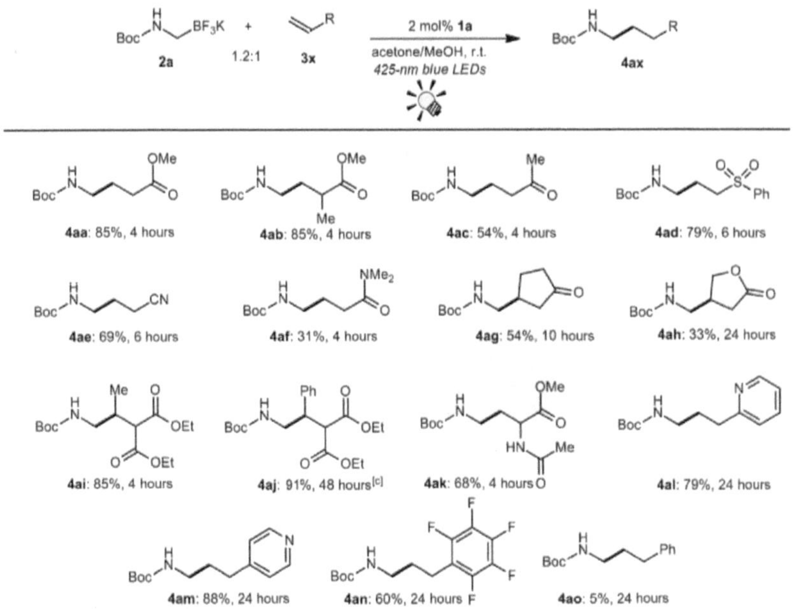

FIGURE 3.23 Scope of the photocatalytic aminomethylation. (Adapted with permission from ref. [119].)

was also conducted on the protective groups that form around the nitrogen atom. A summary of the current photocatalytic aminomethylation of olefins using Boc-protected aminomethyl trifluoroborate can be found in the following **Figure 3.23**.

As a result of the experimental results, a plausible mechanism for the reaction has been presented in **Figure 3.24**. To begin with, the Ir catalyst 1a (Ir[III]) is excited by irradiation of visible light, resulting in the development of *Ir[III]. As evidenced by luminescence quenching experiments, a single electron transfer process occurred between aminomethylborate 2a and the resultant *Ir[III] as a result of the reaction. As a result of the first SET event, aminomethyl radical 2a′ is produced, along with the highly reduced species of Ir[II], by the 1e-oxidation of 2a. A radical intermediate 4′ is produced as a result of the radical addition of 2a′ to olefin 3, which is then reduced via a second single electron transfer event to yield a carbanioinic intermediate. The aminomethylated product 4 is generated after protonation from the solvent, MeOH.

Boronic esters activated with visible light undergo photoredox cross-couplings with C(sp2)-C(sp3) in flow[120]. This is a method for activating boronic esters by means of photoredox reactions. The use of these reagents led to the development of a high-throughput and efficient continuous flow process, which can be used to perform a dual iridium- and nickel-catalyzed

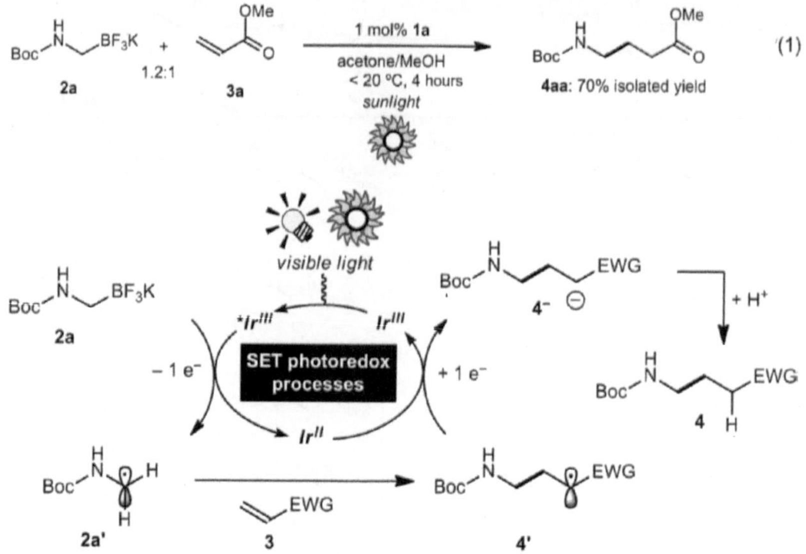

FIGURE 3.24 A plausible reaction mechanism. (Adapted with permission from ref. [119].)

C(sp2)–C(sp3) coupling without having to deal with the solubility issues associated with the potassium trifluoroborate salt. Photoredox activation of the boronic esters requires the formation of an adduct with a Lewis base derived from pyridine. The results of the study led to the formation of a C(sp2)–C(sp3) coupling method that was further simplified by utilizing cyano heteroarenes and boronic esters under flow conditions under visible light.

Figure 3.25 illustrates the scope of the dual Ir/Ni-catalyzed benzyl boronic ester arylation in flow (0.5-mmol scale, 0.1 m in acetone). A high isolated yield of coupled products was associated with organoboron compounds with electron-rich substituents, while compounds with electron-withdrawing substituents were associated with a lower isolated yield of coupled products. It is in accordance with the putative mechanism of single-electron oxidation as the boronates will be more reactive to oxidation when they have a higher electron density. There was an option of varying the aryl bromide coupling partners to allow for the presence of sensitive aldehyde (3h), alkene (3j), and alkyne (3k) groups to be tolerated. The ortho-substituent (3i) is also well tolerated, which is a remarkable fact. A boronic ester consisting of an aryl bromide (3l) was used to verify the orthogonality between the C(sp$_2$) coupling event and the C(sp$_3$) coupling event with the use of boronic esters.

The scope of photoredox arylation of benzyl and allyl boronic esters with cyanoarenes is shown in **Figure 3.26**. In addition to cyanopyridine, this

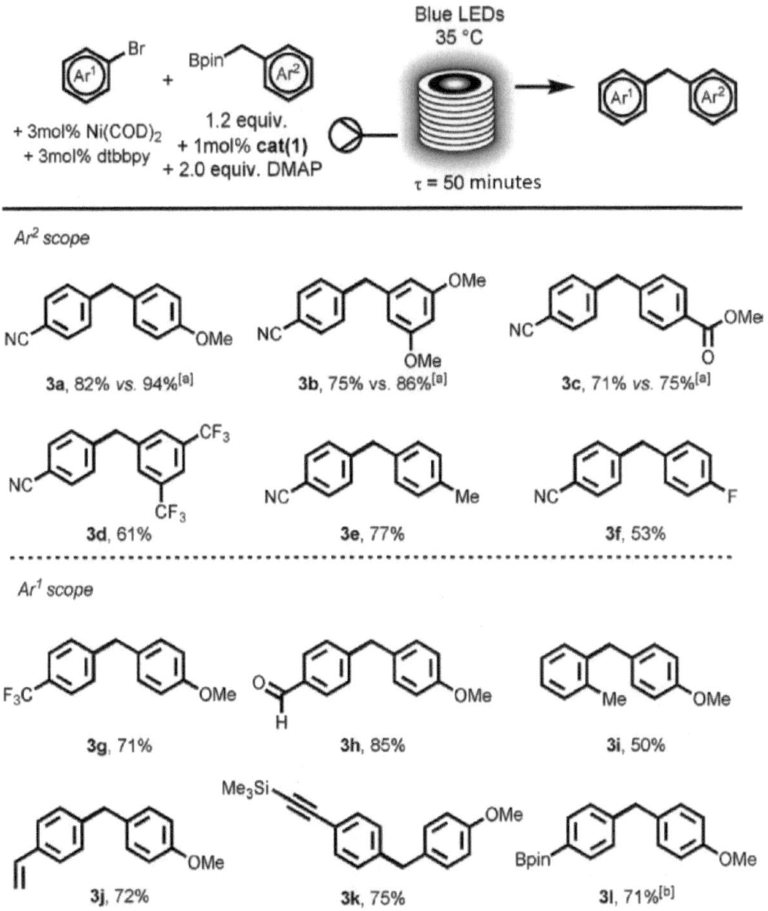

FIGURE 3.25 Scope of the dual Ir/Ni-catalyzed benzyl boronic ester arylation. (Adapted with permission from ref. [120].)

transformation was also successful with several N-heterocycles (**Figure 3.26**). With the 4-cyanopyridine scaffold, it has also been possible to create very interesting coupled products with selective coupling at the most electron-poor 4-position using 4-cyanoquinoline as well as other nitrile substituted derivatives of 4-cyanopyridines. Unprotected 4-cyano-7-azaindole was tolerated by the reaction, which is interesting. With the help of the commercially available allylboronic acid pinacol ester, the standard reaction conditions for allylating N-heterocycles were extended in order to increase its diversity.

For this last protocol to be rationalized, the single-electron reduction and oxidation potentials of Lewis acid–base adducts as well as the equilibrium

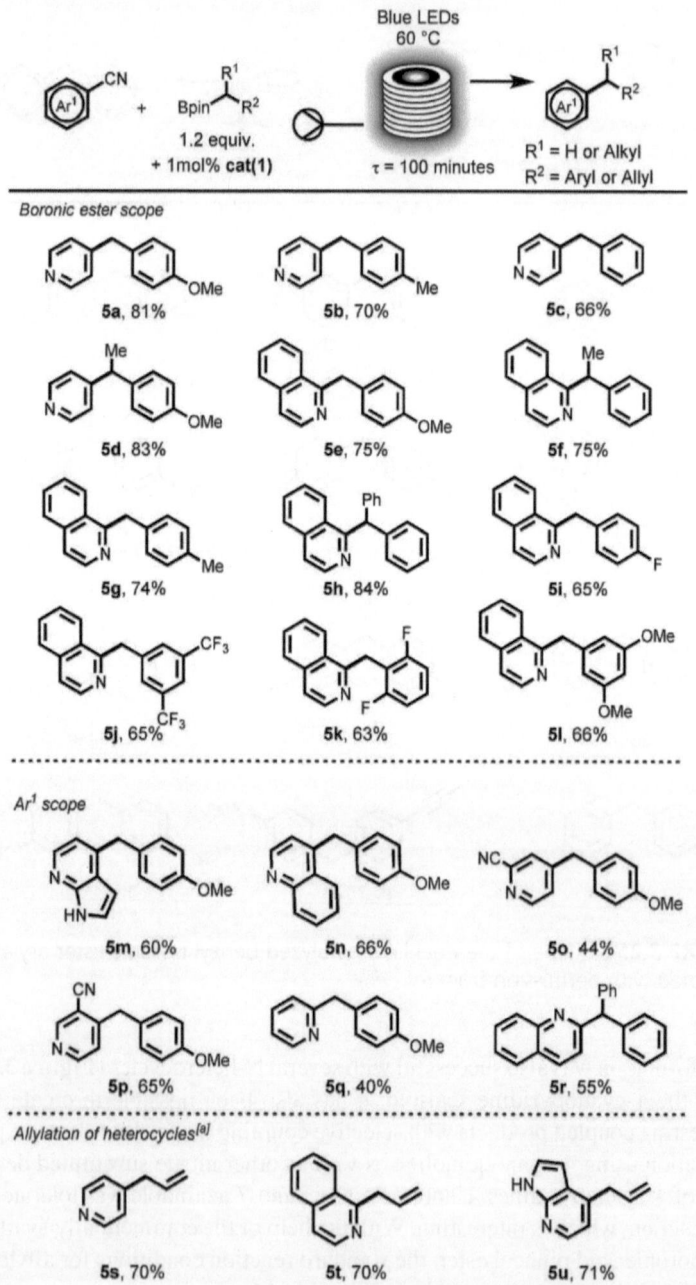

FIGURE 3.26   Arylation of benzyl and allyl boronic esters with cyanoarenes by photoredox. (Adapted with permission from ref. [120].)

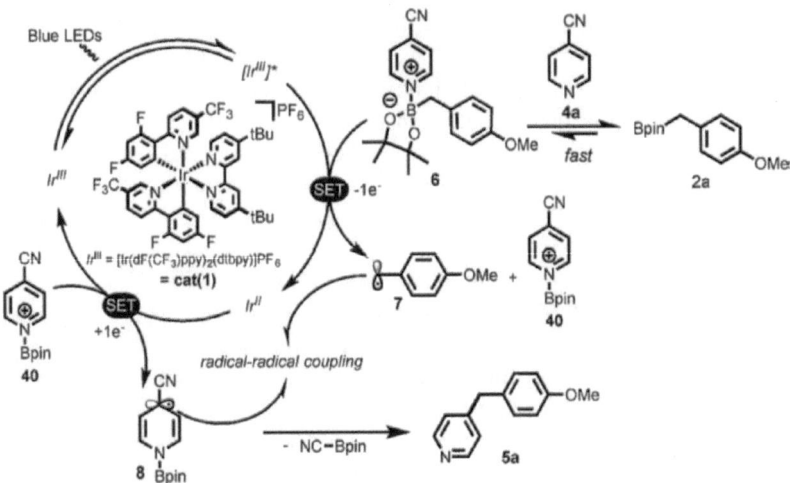

FIGURE 3.27 A mechanistic description of the photoredox coupling between cyanoarenes and organoboronic esters is proposed. (Adapted with permission from ref. [120].)

constants of the adducts have been calculated at the Density Functional Theory (DFT) level. Based on these results, it would appear that the boronic ester and the 4-cyanopyridine are likely to make a complex 6 similar to dimethylaminopyridine (DMAP) (**Figure 3.27**). As a result of the formation of this complex, the single electron oxidation of boronic ester can occur more efficiently. According to this hypothesis, the excited [Ir^{III}]* species is first quenched by 6, which then direct to the rapid cleavage of the C-B bond, resulting in the formation of pyridinium 40 and a carbon-centered radical 7. As a result of the Ir^{II} species being generated at this point, the activated pyridinium were easily reduce by them, generating a radical 8 that couples quickly with 7 to make an intermediate which eliminates the cyanide to produce the coupled product.

The cyclization of alkynes with nitriles to pyridines under the influence of visible-light using Pyrylium Salts as photoredox catalysts has been demonstrated[121]. Using pyrylium salts as photoredox catalysts, regioselective [2 + 2 + 2] cyclizations of aromatic alkynes with nitriles are demonstrated for the preparation of 2,3,6-trisubstituted pyridines under blue LED's illumination. The cycloaddition occurs at room temperature without any metals and as a result of a photooxidative single-electron transfer. It is necessary to use an aromatic alkyne and nitrile mixture to produce the annulation products in good yields.

Researchers have shown that blue LED light can control vinyl ether cationic polymerization[122]. In industrial processes, photoinitiated cationic polymerizations are widely used. Nevertheless, gaining photocontrol over chain growth would greatly increase their utility and allow them to be applied to the design of new complex architectures. There are mild conditions under which the cationic polymerization occurs. As a result of the combination of a metal-free photocatalyst, chain-transfer agent, and light irradiation, a wide variety of poly(vinyl ether)s can be synthesized with good control of their molecular weights and dispersities, as well as excellent chain-end fidelity. Furthermore, photoreversible cation formation in this system allows efficient control of polymer chain growth.

Organic photoredox catalysis mediates a Newman–Kwart rearrangement at ambient temperatures[123]. Among the various methods which have been proposed for synthesizing thiophenols from phenols or blue LEDs, the Newman–Kwart rearrangement stands out as the quintessential method. In spite of that, it has been found that the high temperatures required for the rearrangement of the O-aryl carbamothioates can lead to their decomposition. Using a commercially available organic single-electron photooxidant in catalytic quantities, this study presented a general strategy for the catalysis of O-aryl carbamothioates to S-aryl carbamothioates. A further important aspect of this reaction is the fact that it is facilitated at ambient temperatures.

Aryl amination can be achieved through the use of photoredox catalysis and ligand-free Ni(II) salts[124]. A number of important developments have occurred in the last two decades regarding the use of transition metal catalysts to amination aryl halides to form anilines, a structure that is common to many drug agents, natural product isolates, and fine chemicals, which are common to many natural products. These methods are often able to be highly effective and selective through the design of specialized ligands, which facilitate reductive elimination from a destabilized metal center by providing highly efficient and selective couplings. A general and complementary method could be developed using photoredox catalysis in order to form carbon–nitrogen bonds by destabilizing a metal amido complex, and thus making it possible to use structured ligand systems in a more efficient manner in comparison with conventional catalysis. An article has been published on the development of a distinct mechanistic paradigm for aryl amination involving ligand-free nickel(II) salts, in which a photoredox-catalyzed electron transfer event was used to induce facile reductive elimination from the nickel metal center.

It has been reported that bisindole alkylation of tetrahydroisoquinolines with visible-light induced ring-opening fragmentation led to biologically active bis(indolyl)methane derivatives[125]. It has been demonstrated that visible-light (blue LED) photoredox catalyzed functionalization of tetrahydroisoquinolines with bisindole alkylation has the potential to expedite new avenues for producing bis(indolyl)methane derivatives through ring-opening functionalizations. As part of this process, a series of cascade reactions is performed, including amine oxidation, fragmentation, and Friedel–Craft alkylations as well as photoredox catalysis. Among the bis(indolyl)methane derivatives prepared, five of them contained para substitution groups on the side chain of benzene that inhibited the growth of human breast adenocarcinoma MDA-MB-231 cells, and one of them accomplished significant effects in the migration of these cells. This method and the compounds reported from this investigation have functions like the anticancer molecules reported from our group[126–141].

Brønsted acid-catalysed conjugate addition of photochemically generated α-amino radicals to alkenylpyridines was reported[142]. By the synergistic merger of Brønsted acid and visible light photoredox catalysis, the conjugate addition of α-amino radicals to alkenylpyridines has been accomplished. It was the protonation of the alkenylpyridines, which transiently generated a highly reactive, electrophilic pseudo-iminium ion intermediate, which was key to the development of the reaction. In order to assess the feasibility of an enantioselective catalytic variant, initial investigations have been conducted using chiral phosphoric acids.

3.5.3 Photocatalysis using UV LEDs

The utilization of UV LEDs as irradiators for heterogeneous photocatalysis has already been well established in other applications utilizing solid-state lighting devices for a number of years, but only after they became mainstream alternatives in those applications. In a study conducted by Chen et al., it is claimed to be the first time that a UV LED has been used in heterogeneous photocatalysis[143]. Researchers used a UV LED 375-nm-illuminated photocatalytic reactor to investigate the photo-oxidation of perchloroethylene (PCE). There was a PCE conversion rate of 43% when UV illumination was very weak (at 49 µWcm^{-2}). This study demonstrated that UV LEDs could be used as an alternative source of irradiation for heterogeneous photocatalysis. The authors concluded their study with the prediction that UV LEDs would soon replace conventional UV lamps as an alternative source of irradiation in heterogeneous photocatalysis experiments and applications as a result.

A hybrid photocatalyst consisting of carbon nanofibers (CNF) and zinc oxide was used to synthesize selected aromatic aldehydes under UV-LED illumination[144]. It has been demonstrated that when zinc oxide (ZnO) is prepared by chemical vapor deposition and combined with different amounts of CNF, a hybrid material can be produced that has been thoroughly characterized by a variety of different techniques. In order to determine the catalytic performance of the photocatalyst, the performance of the photocatalyst was analyzed in terms of the synthesis of vanillin (VAD) from vanillyl alcohol (VA). Adding a carbon phase to ZnO was found to be able to increase its surface area as well as the efficiency of its photocatalytic properties by increasing the surface area of the material. It has been attributed to the fact that charge carriers formed on the optical semiconductor are separated efficiently on the basis of their charge. The best-performing material was the one containing 10% CNF in it, which resulted in a 2.5-fold increase in selectivity of the reaction towards vanillin generation when compared to previous experiments, as well as the benefit of being able to carry out the reaction in an aqueous environment. As a result of the use of a similar photocatalyst, the selective synthesis of other aromatic aldehydes, including anisaldehyde, piperonal, and benzaldehyde, has also been successfully achieved. A relationship was proposed between the efficiency of photocatalytic oxidation of alcohols and the aromatic substituents in their rings.

UVA-LED radiation was used to synthesize the pharmaceutical precursor benzhydrol using TiO2-based photocatalysts[145]. With the presence of UVA-LED radiation, an investigation of the effects of several reactional parameters on the photocatalytic reduction of benzophenone into benzhydrol was conducted, which is an important precursor in the pharmaceutical industry. Titanium dioxide was impregnated with noble metals during incipient wetness impregnation. In order to characterize the TiO2-based materials, a variety of techniques have been used. The selectivity of benzhydrols has been shown to be highly influenced by the nature of the solvent and by the presence of KOH in the solvent (which serves as a nucleophilic agent). In terms of efficiency, the most efficient system was obtained using 5 mM of KOH, 2-propanol, and 0.1 g/L of catalyst. After 10 minutes of reaction time, this system was able to achieve 100% selectivity of benzhydrol. In comparison with TiO_2-based materials, the Pd/TiO_2 material showed a high efficiency in benzhydrol formation (in the absence of KOH). In the case of Pd/TiO_2 materials, a decline in photoluminescence intensity was observed, which suggests that electrons were effectively transferred from TiO_2 to Pd nanoparticles during the photoluminescence process.

3.6 CRITICAL ANALYSIS AND FUTURE SCOPE

By examining the reactions shown here critically, it is evident a number of new discoveries have been made. Oxidative-aromatization, for example, has been found to be highly regioselective for triplet energy transfer. It can be seen that the activation of azide occurs through an exceptional mechanistic route, in terms of the mechanism. In order to accomplish this, an energy transfer process is used instead of an electron transfer process. A variety of intramolecular and intermolecular methods have been developed, both of which are very effective. Several cyclobutanones with controlled stereochemistry and sterically hindered structures have been prepared. It has been found that several strategies for the transfer of energy have been developed for the synthesis of polysubstituted azabicyclo compounds, as well as asymmetric synthesis with high enantioselectivity. A sequential method may be suitable for some examples so that the substrates can be coordinated with the chiral catalyst in the most suitable manner in order to achieve the best results.

The use of reactive nitrene, the interaction of hydrogen bonding with enantiopure metal complexes in a photocascade process, and atom-economic methods for C-H amidation and EnT pathway have also been described. In addition, a number of methods have been described that utilize hydrogen bonding with the enantiopure metal complex. A number of examples of single electron transfer processes are also presented in this chapter, which can be used as a method to synthesize new organic compounds under the influence of photo-radiation. Using these procedures, several intellectually challenging methods are available, making this topic extremely valuable. There is a great potential for this method to be used with heterocycles, reactive ring systems, and heteroatomic systems with multiple heteroatoms, as well as carbocycles. There is no doubt that the combination of chiral synthesis, sequential addition, as well as the possibility of redox reactions via energy transfer methods has provided the process with considerable value.

3.7 CONCLUSIONS

In this chapter, it presents a review of some of the most recent developments in photocatalysis that utilize energy-efficient LEDs as irradiation sources for photocatalysis. It is important to choose an LED that has a dominant wavelength that closely matches the absorption spectrum of the photocatalyst in order to maximize the efficiency of the reaction. The advantage of this is that it maximizes the efficiency of the reaction while minimizing unwanted side reactions, which is a huge benefit as it maximizes the reaction efficiency. It is expected that future research in this area may focus on developing inexpensive, less toxic, and more efficient photocatalysts that can be used to synthesize a variety of medicinally important compounds as well as natural products in the future.

ACKNOWLEDGMENTS

AD is grateful to CEA-Grenoble, Joseph Fourier University, University of Göttingen, University of California, Los Angeles, Prince Mohammad Bin Fahd University for their support. BKB is grateful to US NIH, US NCI, Texas Kleberg Foundation, Stevens Institute of Technology, University of Texas M. D. Anderson Cancer Center, University of Texas-Pan American, Community Health Systems of Texas, Prince Mohammad Bin Fahd University for their support.

REFERENCES

1. Lu Z, Yoon TP. Visible light photocatalysis of [2+2] styrene cycloadditions by energy transfer. *Angew Chem.* 2012;124(41):10475–10478.
2. Zou YQ, Duan SW, Meng XG, Hu XQ, Gao S, Chen JR, et al. Visible light induced intermolecular [2+2]-cycloaddition reactions of 3-ylideneoxindoles through energy transfer pathway. *Tetrahedron.* 2012;68(34):6914–6919.
3. Prasad Hari D, Hering T, Koenig B. The photoredox-catalyzed Meerwein addition reaction: intermolecular amino-arylation of alkenes. *Angew Chem Int Ed.* 2014;53(3):725–728.
4. Xi Y, Yi H, Lei A. Synthetic applications of photoredox catalysis with visible light. *Org Biomol Chem.* 2013;11(15):2387–2403.
5. Xuan J, Lu LQ, Chen JR, Xiao WJ. Visible-light-driven photoredox catalysis in the construction of carbocyclic and heterocyclic ring systems. *Eur J Org Chem.* 2013;2013(30):6755–6770.
6. Tucker JW, Stephenson CR. Shining light on photoredox catalysis: theory and synthetic applications. *J Org Chem.* 2012;77(4):1617–1622.
7. Hari DP, König B. Die photokatalytische Meerwein-Arylierung: eine klassische Aryldiazoniumsalz-Reaktion in neuem Licht. *Angew Chem.* 2013; 125(18):4832–4842.
8. Ravelli D, Protti S, Fagnoni M. Carbon–carbon bond forming reactions via photogenerated intermediates. *Chem Rev.* 2016;116(17):9850–9913.
9. Lang X, Zhao J, Chen X. Cooperative photoredox catalysis. *Chem Soc Rev.* 2016;45(11):3026–3038.
10. Yoon TP. Photochemical stereocontrol using tandem photoredox–chiral Lewis acid catalysis. *Acc Chem Res.* 2016;49(10):2307–2315.
11. Narayanam JM, Stephenson CR. Visible light photoredox catalysis: applications in organic synthesis. *Chem Soc Rev.* 2011;40(1):102–113.
12. Chen JR, Hu XQ, Lu LQ, Xiao WJ. Visible light photoredox-controlled reactions of N-radicals and radical ions. *Chem Soc Rev.* 2016;45(8):2044–2056.
13. Shaw MH, Twilton J, MacMillan DW. Photoredox catalysis in organic chemistry. *J Org Chem.* 2016;81(16):6898–6926.

14. Fabry DC, Rueping M. Merging visible light photoredox catalysis with metal catalyzed C–H activations: on the role of oxygen and superoxide ions as oxidants. *Acc Chem Res.* 2016;49(9):1969–1979.
15. Iriondo-Alberdi J, Greaney MF. Photocycloaddition in natural product synthesis. *Eur J Org Chem.* 2007;2007(29):4801–4815.
16. Hoffmann N. Photochemical reactions as key steps in organic synthesis. *Chem Rev.* 2008;108(3):1052–1103.
17. Bach T, Hehn JP. Photochemical reactions as key steps in natural product synthesis. *Angew Chem Int Ed.* 2011;50(5):1000–1045.
18. Nagib DA, MacMillan DW. Trifluoromethylation of arenes and heteroarenes by means of photoredox catalysis. *Nature.* 2011;480(7376):224–228.
19. DiRocco DA, Rovis T. Catalytic asymmetric α-acylation of tertiary amines mediated by a dual catalysis mode: N-heterocyclic carbene and photoredox catalysis. *J Am Chem Soc.* 2012;134(19):8094–8097.
20. Kalyani D, McMurtrey KB, Neufeldt SR, Sanford MS. Room-temperature C–H arylation: merger of Pd-catalyzed C–H functionalization and visible-light photocatalysis. *J Am Chem Soc.* 2011;133(46):18566–18569.
21. Maity S, Zhu M, Shinabery RS, Zheng N. Intermolecular [3+2] cycloaddition of cyclopropylamines with olefins by visible-light photocatalysis. *Angew Chem Int Ed.* 2012;51(1):222–226.
22. Liu Q, Wu LZ. Recent advances in visible-light-driven organic reactions. *Natl Sci Rev.* 2017;4(3):359–380.
23. Marzo L, Pagire SK, Reiser O, König B. Visible-light photocatalysis: does it make a difference in organic synthesis? *Angew Chem Int Ed.* 2018;57(32): 10034–10072.
24. Brimioulle R, Lenhart D, Maturi MM, Bach T. Enantioselective catalysis of photochemical reactions. *Angew Chem Int Ed.* 2015;54(13):3872–3890.
25. Wang Y, Wang N, Zhao J, Sun M, You H, Fang F, et al. Visible-light-promoted site-specific and diverse functionalization of a C(sp3)–C(sp3) bond adjacent to an arene. *ACS Catal.* 2020;10(12):6603–6612. doi:10.1021/acscatal.0c01495
26. Das A. Recent developments in semipolar InGaN laser diodes. *Semiconductors.* 2021;55(2):272–282. doi:10.1134/S106378262102010X
27. Das A. A systematic exploration of InGaN/GaN quantum well-based light emitting diodes on semipolar orientations. *Opt Spectrosc.* 2022;130(3):137–149. doi:10.1134/S0030400X2203002X
28. Bhatkhande DS, Kamble SP, Sawant SB, Pangarkar VG. Photocatalytic and photochemical degradation of nitrobenzene using artificial ultraviolet light. *Chem Eng J.* 2004;102(3):283–290. doi:10.1016/j.cej.2004.05.009
29. Kuo WS, Ho PH. Solar photocatalytic decolorization of methylene blue in water. *Chemosphere.* 2001;45(1):77–83. doi:10.1016/S0045-6535(01)00008-X
30. Gondal MA, Sayeed MN, Seddigi Z. Laser enhanced photo-catalytic removal of phenol from water using p-type NiO semiconductor catalyst. *J Hazard Mater.* 2008;155(1):83–89. doi:10.1016/j.jhazmat.2007.11.066
31. Bergh AA, Dean PJ. Light-emitting diodes. *Proc IEEE.* 1972;60(2):156–223. doi:10.1109/PROC.1972.8592
32. Zhao J, Brosmer JL, Tang Q, Yang Z, Houk KN, Diaconescu PL, et al. Intramolecular crossed [2+2] photocycloaddition through visible light-induced energy transfer. *J Am Chem Soc.* 2017;139(29):9807–9810.

33. Das A, Banik BK. 15 - Versatile Thiosugars in Medicinal Chemistry. In: Banik BK, ed. *Green Approaches in Medicinal Chemistry for Sustainable Drug Design*. Advances in Green and Sustainable Chemistry. Elsevier; 2020:549–574. doi:10.1016/B978-0-12-817592-7.00015-0

34. Das A, Banik BK. Versatile Thiosugars in Medicinal Chemistry. In: Banik BK, ed. *Green Approaches in Medicinal Chemistry for Sustainable Drug Design*. Elsevier; 2024:409–441.

35. Das A, Yadav RN, Banik BK. Ascorbic acid-mediated reactions in organic synthesis. *Curr Organocatalysis*. 2010;7(3):212–241.

36. Das A, Banik BK. Versatile synthesis of organic compounds derived from ascorbic acid. *Curr Organocatalysis*. 2022;9(1):14–33.

37. Das A, Banik BK. Graphene Oxide and Modified Graphene Oxide-Mediated Synthesis of Medicinally Active Compounds. In: Banik BK, ed. *Green Approaches in Medicinal Chemistry for Sustainable Drug Design*. Elsevier; 2024:13–44.

38. Das A, Banik BK. Synthesis of Natural Products by Photochemistry. In: Banik BK, ed. *Green Approaches in Medicinal Chemistry for Sustainable Drug Design*. Elsevier; 2024:259–283.

39. Das A, Banik BK. Green Synthesis of Biologically Active pyrroles and related substrates via C-H Functionalization. In: Banik BK, ed. *Green Approaches in Medicinal Chemistry for Sustainable Drug Design*. Elsevier; 2024:101–131.

40. Das A, Ashraf MW, Banik BK. Thione derivatives as medicinally important compounds. *ChemistrySelect*. 2021;6(34):9069–9100. doi:10.1002/slct.202102398

41. Kawasaki T, Higuchi K. Simple indole alkaloids and those with a nonrearranged monoterpenoid unit. *Nat Prod Rep*. 2005;22(6):761–793.

42. Zhao G, Roudaut C, Gandon V, Alami M, Provot O. Synthesis of 2-substituted indoles through cyclization and demethylation of 2-alkynyldimethylanilines by ethanol. *Green Chem*. 2019;21(15):4204–4210.

43. Bandini M, Eichholzer A. Catalytic functionalization of indoles in a new dimension. *Angew Chem Int Ed*. 2009;48(51):9608–9644.

44. Kochanowska-Karamyan AJ, Hamann MT. Marine indole alkaloids: potential new drug leads for the control of depression and anxiety. *Chem Rev*. 2010;110(8):4489–4497.

45. Zhuo CX, Wu QF, Zhao Q, Xu QL, You SL. Enantioselective functionalization of indoles and pyrroles via an in situ-formed spiro intermediate. *J Am Chem Soc*. 2013;135(22):8169–8172.

46. Angerer EV, Knebel N, Kager M, Ganss B. 1-(Aminoalkyl)-2-phenylindoles as novel pure estrogen antagonists. *J Med Chem*. 1990;33(9):2635–2640.

47. Chu L, Hutchins JE, Weber AE, Lo JL, Yang YT, Cheng K, et al. Initial structure–activity relationship of a novel class of nonpeptidyl GnRH receptor antagonists: 2-arylindoles. *Bioorg Med Chem Lett*. 2001;11(4):509–513.

48. Ambrus JI, Kelso MJ, Bremner JB, Ball AR, Casadei G, Lewis K. Structure–activity relationships of 2-aryl-1H-indole inhibitors of the NorA efflux pump in *Staphylococcus aureus*. *Bioorg Med Chem Lett*. 2008;18(15):4294–4297.

49. Nakhi A, Prasad B, Reddy U, Rao RM, Sandra S, Kapavarapu R, et al. A new route to indoles via in situ desilylation–Sonogashira strategy: identification of novel small molecules as potential anti-tuberculosis agents. *MedChemComm*. 2011;2(10):1006–1010.

50. Willoughby CA, Hutchins SM, Rosauer KG, Dhar MJ, Chapman KT, Chicchi GG, et al. Combinatorial synthesis of 3-(amidoalkyl) and 3-(aminoalkyl)-2-arylindole derivatives: discovery of potent ligands for a variety of G-protein coupled receptors. *Bioorg Med Chem Lett.* 2002;12(1):93–96.
51. Maity S, Zheng N. A visible-light-mediated oxidative C-N bond formation/aromatization cascade: photocatalytic preparation of N-arylindoles. *Angew Chem Int Ed.* 2012;51(38):9562–9566.
52. Das A. LED light sources in organic synthesis: an entry to a novel approach. *Lett Org Chem.* 2022;19(4):283–292.
53. Yadav RN, Hossain F, Das A, Srivastava AK, Banik BK. Organocatalysis: a recent development on stereoselective synthesis of o-glycosides. *Catal Rev.* 2022;66(1):1–118. doi:10.1080/01614940.2022.2041303
54. Das A, Banik BK. Sustainable reactions in the synthesis of heterocycles. *Curr Organocatalysis.* 2022;9(1):3–3. doi:10.2174/2213337209012203281 64523
55. Yadav RN, Shaikh AL, Das A, Ray D, Banik BK. Asymmetric synthesis of 3-pyrrole substituted β-lactams through p-toluene sulphonic acid-catalyzed reaction of azetidine-2,3-diones with hydroxyprolines. *Curr Organocatalysis.* 2022;9(4):337–345.
56. Das A, Yadav RN, Banik BK. A novel Baker's yeast-mediated microwave-induced reduction of racemic 3-keto-2-azetidinones: facile entry to optically active hydroxy β-lactam derivatives. *Curr Organocatalysis.* 2022;9(2):195–198.
57. Das A, Yadav RN, Banik BK. Conceptual design and cost-efficient environmentally benign synthesis of beta-lactams. *Phys Sci Rev.* 2022;8 (11):4053-4084. Published online May 4, 2022. doi:10.1515/psr-2021-0088
58. Das A, Yadav R, Banik BK. 10 Conceptual Design and Cost-Efficient Environmentally Benign Synthesis of Beta-Lactams. In: *10 Conceptual Design and Cost-Efficient Environmentally Benign Synthesis of Betalactams.* De Gruyter; 2022:357–388. doi:10.1515/9783110797428-010
59. Xia X, Xuan J, Wang Q, Lu L, Chen J, Xiao W. Synthesis of 2-Substituted indoles through visible light-induced photocatalytic cyclizations of styryl azides. *Adv Synth Catal.* 2014;356(13):2807–2812.
60. Zhu S, Pathigoolla A, Lowe G, Walsh DA, Cooper M, Lewis W, et al. sulfonylative and azidosulfonylative cyclizations by visible-light-photosensitization of sulfonyl azides in THF. *Chem Weinh Bergstr Ger.* 2017;23(69):17598.
61. Rostoll-Berenguer J, Blay G, Pedro JR, Vila C. 9, 10-phenanthrenedione as visible-light photoredox catalyst: a green methodology for the functionalization of 3, 4-dihydro-1, 4-benzoxazin-2-ones through a Friedel-Crafts reaction. *Catalysts.* 2018;8(12):653.
62. Rogers DA, Brown RG, Brandeburg ZC, Ko EY, Hopkins MD, LeBlanc G, et al. Organic dye-catalyzed, visible-light photoredox bromination of arenes and heteroarenes using N-bromosuccinimide. *ACS Omega.* 2018;3(10):12868–12877.
63. Nishina Y, Ohtani B, Kikushima K. Bromination of hydrocarbons with CBr4, initiated by light-emitting diode irradiation. *Beilstein J Org Chem.* 2013;9:1663–1667. doi: https://doi.org/10.3762/bjoc.9.190
64. Das A, Banik BK. Dipole Moment Studies on Beta Lactams. In: Banik BK, ed. *Green Approaches in Medicinal Chemistry for Sustainable Drug Design.* Elsevier; 2024:523–542.

65. Das A, Banik BK. Dipole moment in medicinal research: green and sustainable approach. In: Banik BK, ed. *Green Approaches in Medicinal Chemistry for Sustainable Drug Design.* Elsevier; 2024:561–602.

66. Das A, Das A, Banik BK. Influence of dipole moments on the medicinal activities of diverse organic compounds. *J Indian Chem Soc.* 2021;98(2):100005. doi:10.1016/j.jics.2021.100005

67. Das A, Banik BK. β-lactams: geometry, dipole moment and anticancer activity. *J Indian Chem Soc.* 2020; 97(11b)(Nov 2020):2461–2467. doi:10.5281/zenodo.5656689

68. Das A, Alqashqari AA, Banik BK. Quantum mechanical calculations of dipole moment of diverse imines. *J Indian Chem Soc.* 2021;97(9b):1563–1566.

69. Das A, Banik BK. Dipole moment studies on α-hydroxy-β-lactam derivatives. *J Indian Chem Soc.* 2021;97(9b):1567–1571.

70. Das A, Banik BK. Dipole moment and anticancer activity of beta lactams. *Indian J Pharm Sci.* 2021;83(5):1071–1074. doi:10.36468/pharmaceutical-sciences.862

71. Das A, Banik BK. Computational studies of physicochemical parameters on optically active anticancer β-lactams. *Heterocycl Lett.* 2023;13(1).

72. Das A, Yadav R, Banik BK. Dipole moment studies on anticancer polyaromatic compounds. *Heterocycl Lett.* 2024;14(2):287–291.

73. Das A, Banik BK. Studies on dipole moment of penicillin isomers and related antibiotics. *J Indian Chem Soc.* 2020;97(6):911–915.

74. Das A, Bose AK, Banik BK. Stereoselective synthesis of β-lactams under diverse conditions: unprecedented observations. *J Indian Chem Soc.* 2020; 97(6):917–925.

75. Das A, Banik BK. 26 - Dipole Moment in Medicinal Research: Green and Sustainable Approach. In. In: Banik BK, ed. *Green Approaches in Medicinal Chemistry for Sustainable Drug Design.* Advances in Green and Sustainable Chemistry. Elsevier; 2020:921–964. doi:10.1016/B978-0-12-817592-7.00021-6

76. Das A, Banik BK. *Microwaves in Chemistry Applications: Fundamentals, Methods and Future Trends.* Elsevier Science; 2021.

77. Das A, Banik BK. Chapter 1 - Foundational Principles of Microwave Chemistry. In: Das A, Banik B, eds. *Microwaves in Chemistry Applications.* Advances in Green and Sustainable Chemistry. Elsevier; 2021:3–26. doi:10.1016/B978-0-12-822895-1.00005-9

78. Das A, Banik BK. Chapter 2 - Microwave Equipment for Chemistry. In: Das A, Banik B, eds. *Microwaves in Chemistry Applications.* Advances in Green and Sustainable Chemistry. Elsevier; 2021:27–59. doi:10.1016/B978-0-12-822895-1.00002-3

79. Das A, Banik BK. Chapter 3 - Modeling and Interpreting Microwave Effects. In: Das A, Banik B, eds. *Microwaves in Chemistry Applications.* Advances in Green and Sustainable Chemistry. Elsevier; 2021:61–104. doi:10.1016/B978-0-12-822895-1.00007-2

80. Das A, Banik BK. Chapter 4 - Microwave-Assisted Synthesis of Oxygen- and sulfur-Containing Organic Compounds. In: Das A, Banik B, eds. *Microwaves in Chemistry Applications.* Advances in Green and Sustainable Chemistry. Elsevier; 2021:107–142. doi:10.1016/B978-0-12-822895-1.00010-2

81. Das A, Banik BK. Chapter 5 - Microwave-Assisted Synthesis of N-Heterocycles. In: Das A, Banik B, eds. *Microwaves in Chemistry Applications.* Advances in Green and Sustainable Chemistry. Elsevier; 2021:143–198. doi:10.1016/B978-0-12-822895-1.00006-0

82. Das A, Banik BK. Chapter 6 - Microwave-Assisted Oxidation and Reduction Reactions. In: Das A, Banik B, eds. *Microwaves in Chemistry Applications. Advances in Green and Sustainable Chemistry.* Elsevier; 2021:199–244. doi:10.1016/B978-0-12-822895-1.00001-1

83. Das A, Banik BK. Chapter 7 - Microwave-Assisted Enzymatic Reactions. In: Das A, Banik B, eds. *Microwaves in Chemistry Applications.* Advances in Green and Sustainable Chemistry. Elsevier; 2021:245–281. doi:10.1016/B978-0-12-822895-1.00009-6

84. Das A, Banik BK. Chapter 8 - Microwave-Assisted Sterilization. In: Das A, Banik B, eds. *Microwaves in Chemistry Applications.* Advances in Green and Sustainable Chemistry. Elsevier; 2021:285–328. doi:10.1016/B978-0-12-822895-1.00011-4

85. Das A, Banik BK. Chapter 9 - Microwave-Assisted CVD Processes for Diamond Synthesis. In: Das A, Banik B, eds. *Microwaves in Chemistry Applications.* Advances in Green and Sustainable Chemistry. Elsevier; 2021:329–374. doi:10.1016/B978-0-12-822895-1.00004-7

86. Das A, Banik BK. Chapter 10 - Future Trends in Microwave Chemistry and Biology. In: Das A, Banik B, eds. *Microwaves in Chemistry Applications.* Advances in Green and Sustainable Chemistry. Elsevier; 2021:375–384. doi:10.1016/B978-0-12-822895-1.00003-5

87. Das A, Yadav RN, Banik BK. Microwave-induced conversion of electromagnetic Energy into heat energy in different solvents: synthesis of β-lactams. *Chem J Mold.* 2022;17(1):62–66. doi:10.19261/cjm.2021.864

88. Das A, Banik BK. Microwave-induced biocatalytic reactions toward medicinally important compounds. *Phys Sci Rev.* 2022;7(4-5):507–538. doi:10.1515/psr-2021-0064

89. Das A, Banik BK 3 Microwave-Induced Biocatalytic Reactions Toward Medicinally Important Compounds. In: *3 Microwave-Induced Biocatalytic Reactions Toward Medicinally Important Compounds.* De Gruyter; 2022:57–88. doi:10.1515/9783110732542-003

90. Das A, Yadav R, Banik B. Microwave-induced surface-mediated highly efficient regioselective nitration of aromatic compounds: effects of penetration depth. *Asian J Chem.* 2021;33:2203–2206. doi:10.14233/ajchem.2021.23131

91. Das A, Banik BK Microwave-induced catalytic transfer hydrogenation in different solvents toward optically active hydroxy beta lactams: effects of penetration depth. *Asian J Org Med Chem.* 2023;8(1):7–10.

92. Das A, Banik BK. Microwave in research – more miracles. *Heterocycl Lett.* 2024;14(2):449–456.

93. Das A, Banik BK. Expeditious synthesis of oxygen and sulfur heterocycles by microwave. *Heterocycl Lett.* 2024;14(2):257–267.

94. Das A, Yadav R, Banik BK. Microwave-induced ferrier rearrangement of hyroxy beta-lactams with glycals. *Appl Chem Eng.* 2024;7(2):1870–1870.

95. Liu Q, Zhu F, Jin X, Wang X, Chen H, Wu L. Visible-light-driven intermolecular [2+2] cycloadditions between coumarin-3-carboxylates and acrylamide analogs. *Chem Eur J.* 2015;21(29):10326–10329.

96. Wang C, Lin Z. Intermolecular [2+2] cycloaddition of 1, 4-dihydropyridines with olefins via energy transfer. *Org Lett.* 2017;19(21):5888–5891.

97. Huang X, Quinn TR, Harms K, Webster RD, Zhang L, Wiest O, et al. Direct visible-light-excited asymmetric Lewis acid catalysis of intermolecular [2+2] photocycloadditions. *J Am Chem Soc.* 2017;139(27):9120–9123.

98. Skubi KL, Kidd JB, Jung H, Guzei IA, Baik MH, Yoon TP. Enantioselective excited-state photoreactions controlled by a chiral hydrogen-bonding iridium sensitizer. *J Am Chem Soc.* 2017;139(47):17186–17192.

99. Das A, Banik BK. Tellurium-based solar cells. *Phys Sci Rev.* 2022;8(12):4631–4658. Published online May 18, 2022. doi:10.1515/psr-2021-0110

100. Das A, Banik BK. 5 Tellurium-Based Solar Cells. In: *5 Tellurium-Based Solar Cells.* De Gruyter; 2022:107–134. doi:10.1515/9783110735840-005

101. Das A, Banik BK Semiconductor characteristics of tellurium and its implementations. *Phys Sci Rev.* 2022;8(12):4659–4687. Published online May 18, 2022. doi:10.1515/psr-2021-0108

102. Das A, Banik BK. 3 Semiconductor Characteristics of tellurium and Its Implementations. In: *3 Semiconductor Characteristics of Tellurium and Its Implementations.* De Gruyter; 2022:55–84. doi:10.1515/9783110735840-003

103. Das A, Das A, Banik BK. Tellurium-based chemical sensors. *Phys Sci Rev.* 2022;8(12):4461-4501. Published online May 17, 2022. doi:10.1515/psr-2021-0116

104. Das A, Das A, Banik BK. 9 Tellurium-Based Chemical Sensors. In: *9 Tellurium-Based Chemical Sensors.* De Gruyter; 2022:183–224. doi:10.1515/9783110735840-009

105. Das A, Ray D, Banik BK. Tellurium in carbohydrate synthesis. *Phys Sci Rev.* 2022;8(11):4157-4178. Published online May 7, 2022. doi:10.1515/psr-2021-0109

106. Das A, Ray D, Banik BK. 4 Tellurium in Carbohydrate Synthesis. In: *4 Tellurium in Carbohydrate Synthesis.* De Gruyter; 2022:85–106. doi:10.1515/9783110735840-004

107. Aldawood SAA, Das A, Banik BK. Tellurium-induced cyclization of olefinic compounds. *Phys Sci Rev.* 2022;8(12):4569–4609. Published online May 17, 2022. doi:10.1515/psr-2021-0119

108. Aldawood SAA, Das A, Banik BK. 11 Tellurium-Induced Cyclization of Olefinic Compounds. In: *11 Tellurium-Induced Cyclization of Olefinic Compounds.* De Gruyter; 2022:249–290. doi:10.1515/9783110735840-011

109. Ray D, Das A, Mazumdar S, Banik BK. Tellurium-induced functional group activation. *Phys Sci Rev.* 2023;8(12):4821–4838. Published online June 2, 2022. doi:10.1515/psr-2021-0221

110. Ray D, Das A, Mazumdar S, Banik BK. 12 Tellurium-Induced Functional Group Activation. In: *12 Tellurium-Induced Functional Group Activation.* De Gruyter; 2022:291–308. doi:10.1515/9783110735840-012

111. Xuan J, Xia X, Zeng T, Feng Z, Chen J, Lu L, et al. Visible-light-induced formal [3+2] cycloaddition for pyrrole synthesis under metal-free conditions. *Angew Chem Int Ed.* 2014;53(22):5653–5656.

112. Farney EP, Yoon TP. Visible-light sensitization of vinyl azides by transition-metal photocatalysis. *Angew Chem Int Ed.* 2014;53(3):793–797.

113. Brachet E, Ghosh T, Ghosh I, König B. Visible light C–H amidation of heteroarenes with benzoyl azides. *Chem Sci.* 2015;6(2):987–992.

114. Scholz SO, Farney EP, Kim S, Bates DM, Yoon TP. Spin-selective generation of triplet nitrenes: olefin aziridination through visible-light photosensitization of azidoformates. *Angew Chem.* 2016;128(6):2279–2282.

115. Zhang C, Li S, Bureš F, Lee R, Ye X, Jiang Z. Visible light photocatalytic aerobic oxygenation of indoles and pH as a chemoselective switch. *ACS Catal.* 2016;6(10):6853–6860.

116. Yoo WJ, Tsukamoto T, Kobayashi S. Visible light-mediated Ullmann-type C–N coupling reactions of carbazole derivatives and aryl iodides. *Org Lett.* 2015;17(14):3640–3642.

117. Münster N, Parker NA, van Dijk L, Paton RS, Smith MD. Visible light photocatalysis of 6π heterocyclization. *Angew Chem Int Ed.* 2017;56(32): 9468–9472.

118. Zhu S, Das A, Bui L, Zhou H, Curran DP, Rueping M. Oxygen switch in visible-light photoredox catalysis: radical additions and cyclizations and unexpected C–C-bond cleavage reactions. *J Am Chem Soc.* 2013;135(5):1823–1829.

119. Miyazawa K, Koike T, Akita M. Hydroaminomethylation of olefins with aminomethyltrifluoroborate by photoredox catalysis. *Adv Synth Catal.* 2014; 356(13):2749–2755. doi:10.1002/adsc.201400556

120. Lima F, Kabeshov MA, Tran DN, Battilocchio C, Sedelmeier J, Sedelmeier G, et al. Visible light activation of boronic esters enables efficient photoredox C(sp2)–C(sp3) cross-couplings in flow. *Angew Chem Int Ed.* 2016;55(45): 14085–14089. doi:10.1002/anie.201605548

121. Wang K, Meng LG, Wang L. Visible-light-promoted [2 + 2 + 2] cyclization of alkynes with nitriles to pyridines using pyrylium salts as photoredox catalysts. *Org Lett.* 2017;19(8):1958–1961. doi:10.1021/acs.orglett.7b00292

122. Kottisch V, Michaudel Q, Fors BP. Cationic polymerization of vinyl ethers controlled by visible light. *J Am Chem Soc.* 2016;138(48):15535–15538. doi:10.1021/jacs.6b10150

123. Perkowski AJ, Cruz CL, Nicewicz DA. Ambient-temperature Newman–Kwart rearrangement mediated by organic photoredox catalysis. *J Am Chem Soc.* 2015;137(50):15684–15687. doi:10.1021/jacs.5b11800

124. Corcoran EB, Pirnot MT, Lin S, Dreher SD, DiRocco DA, Davies IW, et al. Aryl amination using ligand-free Ni(II) salts and photoredox catalysis. *Science.* 2016;353(6296):279–283. doi:10.1126/science.aag0209

125. Chen CC, Hong BC, Li WS, Chang TT, Lee GH. Synthesis of biologically active bis(indolyl)methane derivatives by bisindole alkylation of tetrahydroisoquinolines with visible-light induced ring-opening fragmentation. *Asian J Org Chem.* 2017;6(4):426–431. doi:10.1002/ajoc.201600415

126. Das A, Banik BK. Advances in heterocycles as DNA intercalating cancer drugs. *Phys Sci Rev.* 2023;8(9):2473–2521. Published online January 5, 2022. doi:10.1515/psr-2021-0065

127. Das A, Banik BK. 4 Advances in Heterocycles as DNA Intercalating Cancer Drugs. In: *Heterocyclic Anticancer Agents.* De Gruyter; 2022:111–160. doi:10.1515/9783110735772-004

128. Banik BK, Das A. *Natural Products as Anticancer Agents.* Elsevier Science; 2023.

129. Banik BK, Das A. Anticancer Activity of Natural Compounds from Marine Plants. In: Banik BK, Das A, eds. *Natural Products as Anticancer Agents.* Elsevier; 2024:237–284.

130. Banik BK, Das A. Anticancer Activity of Natural Compounds from Bacteria. In: Banik BK, Das A, eds. *Natural Products as Anticancer Agents.* Elsevier; 2024:287–328.

131. Banik BK, Das A. Anticancer Activity of Natural Compounds from Fungi. In: Banik BK, Das A, eds. *Natural Products as Anticancer Agents.* Elsevier; 2024:329–366.

132. Banik BK, Das A. Anticancer Drugs from Hormones and Vitamins. In: Banik BK, Das A, eds. *Natural Products as Anticancer Agents*. Elsevier; 2024:369–414

133. Banik BK, Das A. Future Prospect in Anticancer Natural Products. In: Banik BK, Das A, eds. *Natural Products as Anticancer Agents*. Elsevier; 2024:415–426

134. Das A, Banik BK. Anticancer Activity of Natural Compounds from Leaves of the Plants. In: Banik BK, Das A, eds. *Natural Products as Anticancer Agents*. Elsevier; 2024:3–48.

135. Das A, Banik BK. Anticancer Activity of Natural Compounds from Stems/ Barks of the Plants. In: Banik BK, Das A, eds. *Natural Products as Anticancer Agents*. Elsevier; 2024:49–86.

136. Das A, Banik BK. Anticancer Activity of Natural Compounds from Roots of the Plants. In: Banik BK, Das A, eds. *Natural Products as Anticancer Agents*. Elsevier; 2024:87–132.

137. Das A, Banik BK. Anticancer Activity of Natural Compounds from Fruits and Vegetables. In: Banik BK, Das A, eds. *Natural Products as Anticancer Agents*. Elsevier; 2024:133–178.

138. Das A, Banik BK. Anticancer Activity of Natural Compounds from Marine Animals. In: Banik BK, Das A, eds. *Natural Products as Anticancer Agents*. Elsevier; 2024:181–236.

139. Das A, Banik BK. Combatting the Coronavirus Utilizing Natural Cinnamon and Its Derived Products. *Asian Journal of Synthetic & Natural Product Chemistry*. 2023;1(1):11–15.

140. Das A. Quantitative structure-property relationships of taxol, taxotere and their epi-isomers. *J Indian Chem Soc*. 2020;97(11b):2468–2476.

141. Shaikh AL, Das A, Banik BK. Indium-mediated reduction of aromatic nitro groups in β-lactams to oxazines. *Heterocycl Lett*. 2024;14(2):267–272.

142. Hepburn HB, Melchiorre P. Brønsted acid-catalysed conjugate addition of photochemically generated α-amino radicals to alkenylpyridines. *Chem Commun*. 2016;52(17):3520–3523. doi:10.1039/C5CC10401G

143. Chen DH, Ye X, Li K. Oxidation of PCE with a UV LED photocatalytic reactor. *Chem Eng Technol*. 2005;28(1):95–97. doi:10.1002/ceat.200407012

144. Fernandes RA, Sampaio MJ, Da Silva ES, Serp P, Faria JL, Silva CG. Synthesis of selected aromatic aldehydes under UV-LED irradiation over a hybrid photocatalyst of carbon nanofibers and zinc oxide. *Catal Today*. 2019;328:286–292. doi:10.1016/j.cattod.2018.10.061

145. Lopes JC, Sampaio MJ, Rosa B, Lima MJ, Faria JL, Silva CG. Role of TiO2-based photocatalysts on the synthesis of the pharmaceutical precursor benzhydrol by UVA-LED radiation. *J Photochem Photobiol Chem*. 2020;391:112350. doi:10.1016/j.jphotochem.2019.112350

Photo-Mediated Drug-Free Sustainable Therapeutic Approaches

4

A Leap in the Future for Cancer Treatment

Bimal Krishna Banik and Aparna Das

4.1 INTRODUCTION

The global incidence of cancer continues to rise, with more than 10 million new cases being reported each year, making it one of the top causes of death in the world[1]. As a matter of fact, the conventional treatment of cancer is limited to chemotherapy, radiotherapy, and surgery at the moment. In most cases, chemotherapy and radiotherapy are not only harmful to normal cells, but they also result in serious side effects, such as neuropathy, neutropenia, and kidney failure, which may be serious in some cases[2]. Even though surgery

DOI: 10.1201/9781003634249-4

is capable of removing primary tumors and visible metastases, it is often limited in its effectiveness by the propensity of tumors to invade adjacent tissues or to spread to distant sites by micrometastasis[3]. Nanotechnology has significantly contributed to advances in cancer treatments. There have been a number of controversial issues regarding nanoparticles (NPs), including their in vivo cytotoxicity as well as their biosecurity, especially the unexpected accumulation of nanomaterials in the body, which has been investigated in the context of biomedical research[4].

A high degree of spatial selectivity and inherent non-invasiveness of phototherapy methods have demonstrated that they can target tumors precisely and minimize damage to normal cells as well[5]. These advantages have led to widespread use of phototherapy methods in preclinical and clinical cancer treatment. In phototherapy, there are two main types of phototherapy: photodynamic therapy (PDT) as well as photothermal therapy (PTT)[6]. Several reactions can be catalyzed by using different types of light sources and photocatalysts[7]. Lamps, lasers[8], or light-emitting diodes (LEDs)[9] have been utilized when artificial light sources have been preferred over natural light in photocatalysis.

Using photocatalysts to treat cancer is part of the new technique of PDT. There are three main components that are involved in the PDT process, and they are oxygen, light, and photosensitizers (light-responsive materials). It has been demonstrated that hybrid semiconductor photocatalysts absorb light energy, transfer it to molecular oxygen, and then generate cytotoxic ROS (reactive oxygen species)[10]. A few natural products synthesized by photochemical method also has the potential to produce reactive oxygen[11].

It is due to the synergistic combination of inorganic materials with unique physical properties and the targeting capability of biomolecules that nanophotocatalysts have an advantage over conventional photocatalysts. Their multi-functional drug molecules, together with the ideal therapeutic process, which in turn makes them an ideal therapeutic system.[12]. Furthermore, nanophotocatalysts are capable of passing through biological barriers, including the blood-brain barrier.

Despite many nanocarriers being developed to improve these properties to some extent, drug therapy will always involve toxic side effects and limited therapeutic efficacy due to drug content. The potential for drug-free therapeutics to maximize the benefit of the treatment and thus is well worth anticipating, as it is defined as a therapeutic methodology without the use of drugs, without the consumption of therapeutic agents during treatment, yet with an inexhaustible therapeutic capability, thus making it a method of treatment that is worth using. In the past few years, researchers have conducted a lot of research in the hopes of achieving the goal of drug-free therapy. During this chapter, we will discuss the recent advances in photo-mediated drug-free sustainable therapeutic approaches for the treatment of cancer.

4.2 CANCER THERAPY USING NANOPHOTOCATALYSTS

We have been engaged in the synthesis and biological evaluation of diverse organic molecules as anticancer agents[13-27]. To enhance drug efficacy and reduce side effects, nanomaterials are being developed to transport and encapsulate drug molecules in a way that delivers them to diseased sites in a targeted manner[28,29]. All efforts are made to ensure that a high level of drug loading capacity is achieved along the way, but it is also limited. This problem can be resolved better by considering a drug-free approach to cancer therapy as an alternative to dealing with it.

Researchers have reported a calcification procedure to eliminate cancer cells using folic acid and Ca^{2+} injections into cancer sites. These two therapeutic agents, however, were consumed in cancer treatment[30]. Another group has proposed using nanocatalytic medicine to improve cancer therapy, which would lead to drug-free therapies in principle, without the toxic side effects that come with them[31,32]. The majority of catalytic nanomedicines, however, are loaded with toxic drugs and agents that are easy to exhaust. In a drug-free therapy, the development of nanocatalytic nanomedicine is quite intriguing and challenging at the same time. In fact, the application of catalytic methods in the development of drugs or drugs-like substances has been investigated by our group[7,11,33-42].

In order to maintain a drug-free therapy of cancer, it is important to develop drug-free nanocatalysts that are capable of consuming intratumoral endogenous substrates in order to create therapeutic species. Photocatalysis with nanocatalysts for cell proliferation and drug-free cancer treatment has been reported in the literature[43]. A drug-free therapeutic approach has been defined and presented in this paper as a form of therapeutic methodology that does not require the use of traditional toxic drugs to perform the treatment, nor do they require the consumption of therapeutic agents to achieve the desired results, but it does have an inexhaustible therapeutic capability capable of maximizing the therapeutic benefit for the patient. In addition, a Z-scheme $SnS_{1.68}$-$WO_{2.41}$ nanocatalyst has been developed in order to achieve near-infrared (NIR) photocatalytic generation of oxidative holes and hydrogen molecules, which will allow the combination of hole/hydrogen therapy by a drug-free treatment strategy (**Figure 4.1**).

During the NIR irradiation of this nanocatalyst, it was found that this nanocatalyst was capable of oxidizing and consuming intratumoral

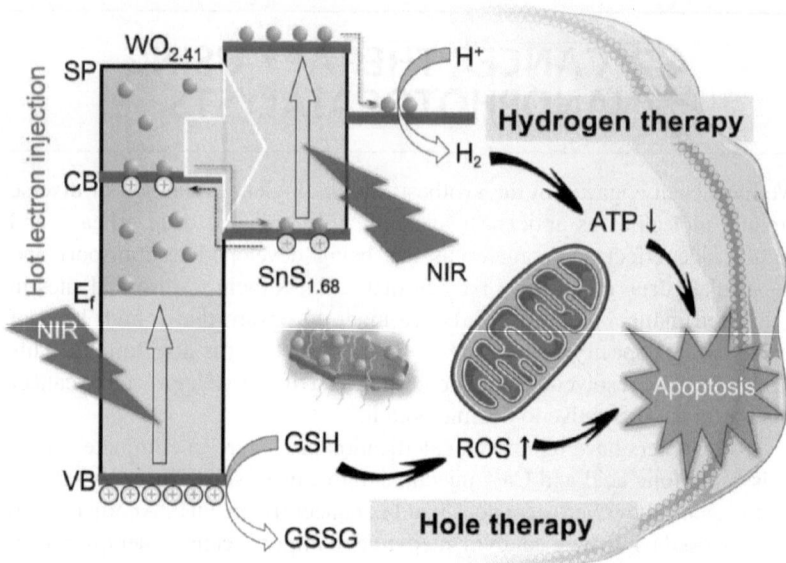

FIGURE 4.1   Schematic illustration of combined hole/hydrogen therapy strategy and mechanisms with the NIR-activable Z-scheme $SnS_{1.68}$–$WO_{2.41}$ nanocatalyst. (Adapted with permission from ref. [43].)

glutathione (GSH) by holes without the aid of any drugs or other therapeutic agents, and simultaneously, it was capable of generating hydrogen molecules in a lasting and controlled manner. On a mechanical level, hydrogen molecules generated in cancer cells and the consumption of GSH destroy intracellular redox balance and inhibit cancer cell energy, respectively, so that they synergistically harm DNA and induce cancer cell apoptosis. The efficacy and mechanism of in vitro hole/hydrogen treatment is depicted in **Figure 4.2**. Using the nanocatalyst, the combined hole/hydrogen treatment of cancer cells is achieved with high levels of efficacy and biosafety. By utilizing a drug-free therapeutic strategy based on catalysis, it is possible for cancer treatment to be achieved with high efficacy and low toxicity. The **Figure 4.3** shows the outcomes of the hole/hydrogen therapy performed in vivo.

For tumor marker detection, using ZnO/AgI nanophotocatalyst, a photochromic immunoassay was recently reported[44]. In the first step, ZnO/AgI NPs were prepared and characterized by TEM (transmission electron microscopy), SEM (scanning electron microscopy), XRD (X-ray powder

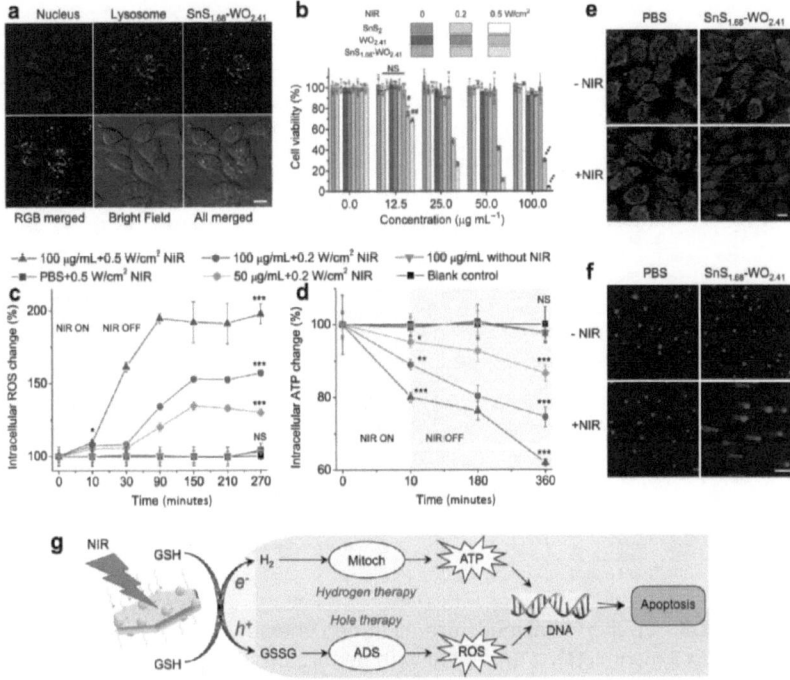

FIGURE 4.2   Combined hole/hydrogen therapy performances and mechanisms of the $SnS_{1.68}$–$WO_{2.41}$ nanocatalyst. (Adapted with permission from ref. [43].)

diffraction), and FTIR (Fourier transform infrared spectrometry). The color development is caused by oxidative reaction between the ZnO/AgI nano-materials and tetramethyl benzidine (TMB) solution. As a result of visible light irradiation, electron transitions occur in ZnO/AgI nanomaterials, which produce photogenerated holes and oxidize TMB to blue solution. By separating photogenerated electrons from holes as a result of the appropriate band width between ZnO and AgI, the oxidation process is more efficient. ZnO/AgI nanomaterial was used as labels in an immunoassay that was based on a sandwich design. It has been found that the absorbance of the reaction solution at 650 nm is positively correlated with the concentration of the antigen. The developed immunoassay showed good performance in terms of detection of CEA (carcinoma embryonic antigen) in the range of 0.1–7.0 ng/mL and the limit of detection was 65 ng/mL. A photochromic immunoassay is also highly selective, repeatable, and stable in comparison to other immunoassays.

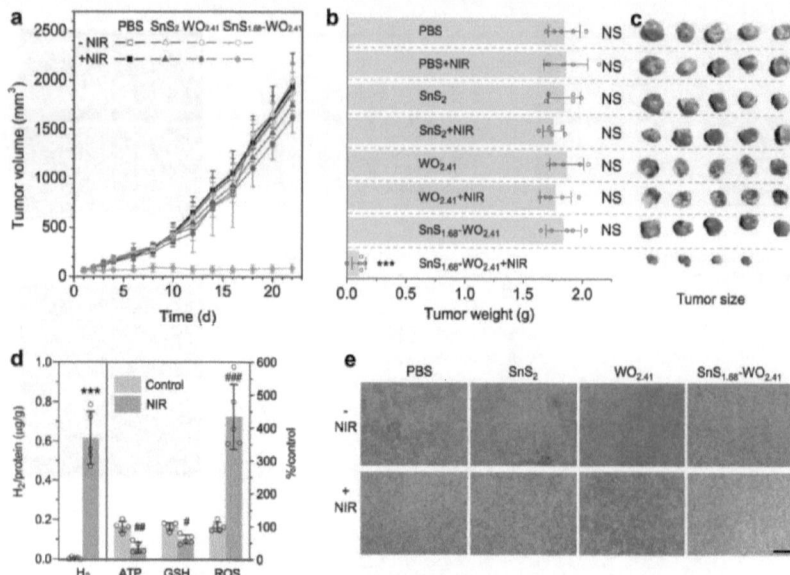

FIGURE 4.3 In vivo combined hole/hydrogen therapy performances of the $SnS_{1.68}$–$WO_{2.41}$ nanocatalyst. (Adapted with permission from ref. [43].)

4.3 CANCER THERAPY USING NPs

The majority of oral malignancies arise from OSCCs (oral squamous cell carcinomas), and most OSCCs develop from precancerous lesions in the mouth, collectively known as OPMDs (potentially malignant disorders)[45]. OLK (Oral leukoplakia) is considered by the WHO (World Health Organization) to be one of the most common OPMDs. OLK is defined as a white impervious, nonscratchable plaque that has the potential to be a cancerous growth when left untreated[46].

An investigation has been published in which a multifunctional photodynamic and photothermal nanoagent was used to treat oral leukoplakia[47]. As a result of the possibility of oral leukoplakia's malignant changes, oral leukoplakia has received a great deal of attention. Even though PSs (photo sensitizers) play an essential role in PDT for OLK, their poor light sensitivity impedes their clinical application.

There has been development of an organic photosensitive ITIC-Th NPs that can be used in oral leukoplakia PTT (photodynamic/photothermal

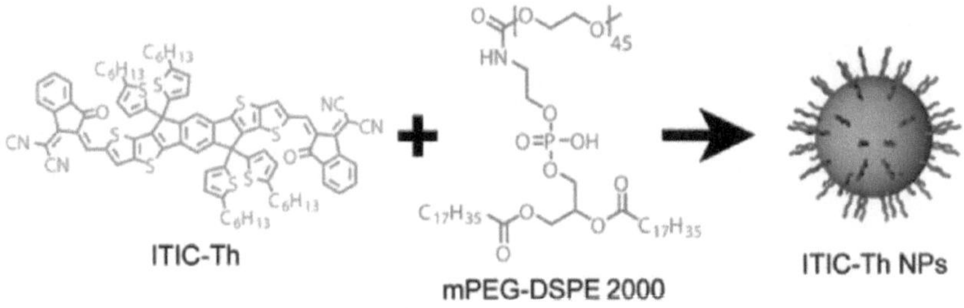

FIGURE 4.4   The illustration of the synthesis of ITIC-Th NPs. (Adapted with permission from ref. [47].)

therapy). The ITIC-Th NPs (**Figure 4.4**) provide both high photothermal conversion efficiency of around 40% and good ROS creation capability under laser irradiation (660 nm), thus making them a very good candidate. Introducing ITIC-Th NPs into oral precancerous animal models that were induced with 4-nitroquinoline 1-oxide (4NQO) successfully suppressed OLK cancerization in vivo without any apparent systemic or topical toxicity (**Figure 4.5**). There is a promising therapeutic strategy that can be applied for PTT and PDT in the treatment of OLK, and this is one of the important interdisciplinary research on the topic of multimodal treatment for Oral leukoplakia that has been conducted (**Figure 4.6**).

In the fight against cancer, the ultimate goal is to develop an effective, nontoxic, tumor-specific immunotherapy that can be administered to patients with metastatic cancer. There has been considerable evidence that checkpoint blockade-based immunotherapies are exceptionally effective, but they only benefit a small percentage of patients whose tumors have already been infiltrated by T cells prior to the immunotherapy. It has been demonstrated that nontoxic NPs mediating PDT are synergistic with immuno-checkpoint blockade to cause antitumor immunity and to counteract the progression of breast cancer[48]. The purpose of this study was to find out whether Zn-pyrophosphate NPs loaded with pyrolipid (ZnP@pyro), a photosensitive photosensitive compound, can kill tumor cells when they are exposed to light by causing apoptosis and necrosis directly, or indirectly, by disrupting tumor vasculature and enhancing the immunogenicity of the tumors.

Furthermore, immunogenic ZnP@pyro PDT treatment sensitizes tumors to checkpoint inhibition through PD-L1 antibodies, which in addition to eliminating the primary 4T1 breast tumor also significantly reduces lung metastasis. ZnP@pyro PDT coupled with anti-PD-L1 treatment demonstrated abscopal effects in both 4T1 and TUBO bilateral syngeneic mice. By

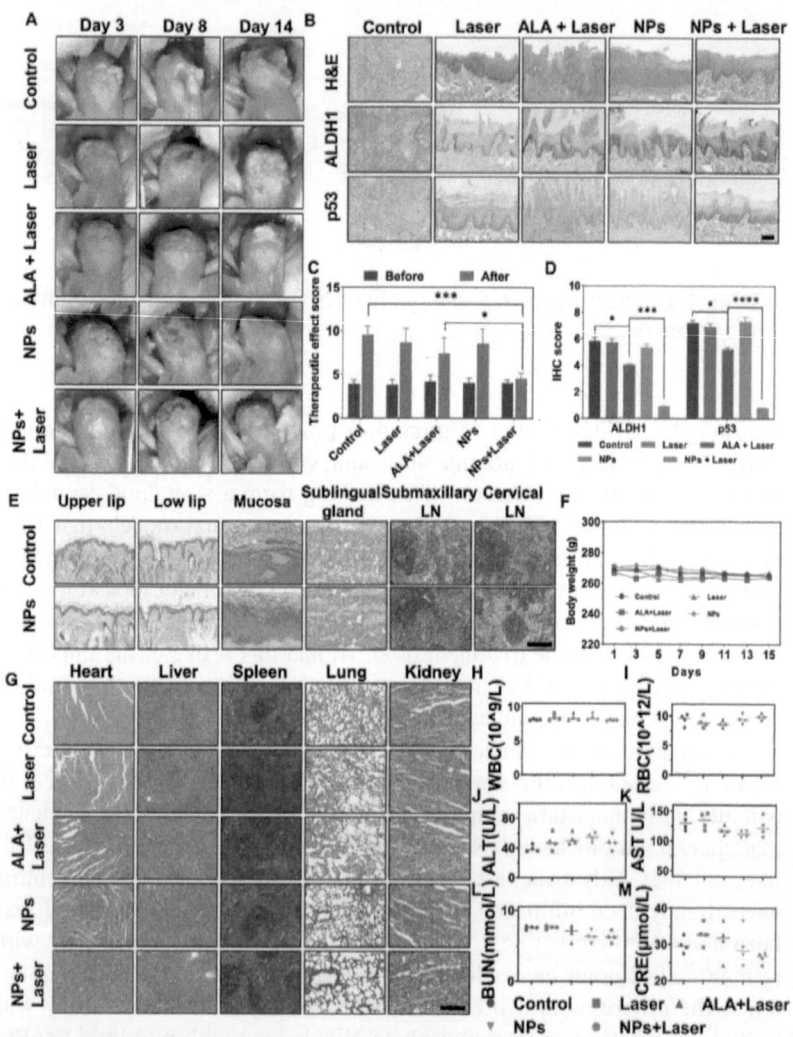

FIGURE 4.5 ITIC-Th NPs therapy intercepts the cancerous progression of oral leukoplakia in 4NQO rats. (Adapted with permission from ref. [47].)

stimulating tumor-specific cytotoxic T cells in a systemic manner, ZnP@pyro PDT leads to the complete elimination of light-irradiated primary tumors as well as the complete inhibition of untreated distant tumors. By activating both the innate and adaptive immune systems in the tumor microenvironment (TME), NP-based PDT appears to enhance the efficacy of checkpoint blockade immunotherapies.

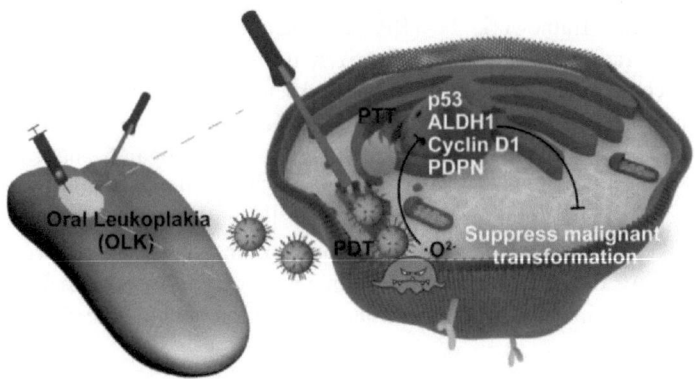

FIGURE 4.6   The illustration of the synergistic PDT/PTT of ITIC-Th NPs after local injection in vivo. (Adapted with permission from ref. [47].)

It is widely recognized that photoactivated therapy, such as PDT and PTT, is a spatiotemporally precise, controllable, and non-invasive method for the treatment of tumors and as a result, it has been attracting a lot of attention in recent years. In spite of this, there is still a challenge in finding highly efficient therapeutic photoactive agents (PAAs) and delivering them into tumors, especially in the core of solid tumors, that are highly targetable. The design and testing of an engineered cell-assisted photoactive NP delivery system for synergistic image-guided photothermal/photodynamic cancer treatment was carried out[49]. An engineered monocyte (MNC)-based PAA system was developed in this study, which is capable of ensuring highly precise and efficient tumor diagnosis and treatment.

The first step was to design and fabricate an NIR emissive PAA molecule with an high photothermal conversion efficiency as well as the strong ability to produce singlet oxygen ($^1O_2$). The PAA molecule was then precisely designed for simultaneous PDT and PTT of tumors and was further fabricated to produce PAA NPs. The PAA NPs were loaded into MNCs and then the MNCs were decorated with cyclic Arg-Gly-Asp (cRGD) groups in order to further improve their ability to target and homing into the deep regions of tumors. In vitro and in vivo, the use of this strategy led to highly efficient solid tumor ablation results, indicating that it has the potential to be used for the treatment of solid tumors in the future.

While immunotherapy has emerged as an effective cancer treatment, only a few cancer patients are able to benefit from it. Cancer immunotherapy can be primed with immunogenic PDT, but tumor hypoxia limits its efficacy. The use of a nanoscale metal-organic framework has been reported to overcome hypoxia during PDT priming of cancer immunotherapy[50]. A nanoscale metal–organic framework, Fe-TBP, has been successfully used in cancer immunotherapy as a nanophotosensitizer, the study demonstrated that the Fe-TBP could overcome tumor hypoxia and sensitize effective PDT, enabling the cancer immune system to target non-inflamed tumors. In both normoxic and hypoxic conditions, Fe-TBP, made up of porphyrin ligands and iron-oxo clusters, sensitized PDT. Fe-TBP mediated PDT significantly improved the efficacy of α-PD-L1 (anti-programmed death-ligand 1) treatment and elicited abscopal effects in a mouse model of colorectal cancer, resulting in over 90% regression of tumors. Mechanistic studies have demonstrated that Fe-TBP-mediated PDT induces significant tumor infiltration of cytotoxic T cells as a result of PDT.

4.4 CANCER THERAPY USING HETEROJUNCTION PHOTOCATALYSTS

With the help of ternary GO-C_3N_4-$AgBr$ heterojunction nanophotocatalysts, colorimetric tumor marker detection could be accomplished using visible light[51]. Visible light was used as an activator in this study to enable a colorimetric immunoassay to detect tumor markers. To start with, the synthesized photocatalyst GO-C_3N_4-$AgBr$ was able to oxidize TMB due to the photogenerated hole, with the oxidation ability being enhanced by the ternary heterojunction structure of the catalyst. Our group has been conducting research on the activation of tellurium by light-mediated exposure and demonstrating the multi-faced behavior of this catalyst as well[52–63].

In contrast, sandwich-type immunocomplexes were constructed using magnetic beads, which were easy to clean as well as easy to separate, depending on the specific binding of antigen and antibody. To complete the immunocomplex, TMB was added to the ABS buffer solution. The color of the sample changed within a few seconds when exposed to visible light radiation. Under optimal conditions, there was a linear relation between absorbance and target protein concentration in the range of 0.5–25 ng mL^{-1} with the detection limit of 20 pg mL^{-1}. Furthermore, the developed colorimetric immunoassay displayed better selectivity, repeatability, and stability as compared to other immunoassays.

In addition to having an enhanced permeability and retention effect, inorganic biophotocatalysts with a suitable size can also be used to Make ROS via a type-I reaction in order to overcome tumor hypoxia via PDT. $BiOI$/$BiOIO_3$ heterojunction biophotocatalyst has been synthesized using a reduction surfactant-assisted one-step method for enhanced photodynamic theranostics overcoming tumor hypoxia[64]. During the study, a facile one-step hydrothermal method was developed to prepare biocompatible BB NCs ($BiOI/BiOIO_3$ heterostructure nanocomposites) by introducing reductive PTMP-PMAA (trithiol-terminated poly-(methacrylic acid)) as the surfactant, which can reduce partial KIO_3 (raw material) into KI, resulting in the production of $BiOIO_3$ and BiOI in one nanocomposite.

Under irradiation with a laser (650 nm), through the generation of photogenerated holes/electrons within the agent, hydroxyl radicals (OH) and singlet oxygen are effectively produced as a result of a type I and type II PDT process, respectively, under both normoxic and hypoxic conditions. This can result in a reduction in mitochondrial membrane potentials (MMP), inhibiting the production of ATP, resulting in HeLa cells dying and eventually causing the tumor to be completely destroyed at a low dose as a result. Furthermore, the agent exhibits extraordinary computed tomography imaging capabilities due to its high X-ray absorption coefficient. This study results in the development of a hypoxic tumor theranostic agent that has a robust biowindow response.

In the process of PDT, oxygen is deficient within tumors, and electron-hole separation in photosensitizers is inefficient, particularly the long-range diffusion of oxygen towards photosensitizers. Hypoxic tumor therapy using Z-scheme heterostructures shows spatiotemporally synchronized oxygen self-supply and ROS production[65]. A novel bismuth sulfide (Bi_2S_3)@bismuth (Bi) Z-scheme heterostructured nanorod (NR) has been developed for hypoxic tumor therapy in which it can deliver oxygen self-supply and produce ROS over a spatiotemporally synchronous time period.

With an NIR laser, both narrow-bandgap Bi and Bi_2S_3 components can be excited to create holes and electrons in abundance. The Z-scheme heterostructure endows Bi_2S_3@Bi NRs with an efficient electron–hole separation ability and potent redox potentials, where the hole on the valence band of Bi_2S_3 can react with water to supply oxygen for the electron on the conduction band of Bi to make ROS. The NR structure showed a promising phototherapeutic effect, which can be utilized during the PDT process to eliminate hypoxic tumors in hypoxic environments by overcoming the obstacles of conventional photosensitizers.

4.5 CANCER THERAPY USING NANOCOMPLEXES

There is a limitation to the effectiveness of conventional light-driven cancer therapeutics, as they require oxygen and visible light to indirectly damage biomolecules, which is the case in deep, hypoxic tumors. Recently, it has been reported that NIR-activated anticancer platinum(IV) complexes have the ability to directly photooxidize biomolecules in a manner that is oxygen independent[66]. By using NIR-activated small molecules of Pt(IV) photooxidants, it is possible to directly oxidize intracellular biomolecules in an oxygen-independent manner, enabling controllable and effective elimination of cancer stem cells (CSCs). There is low toxicity in the dark with these Pt(IV) complexes accumulating in endoplasmic reticulum.

The resultant metal-enhanced photooxidation effect occurs as a consequence of the irradiation process, which causes them to strongly photooxidize some survival-related biomolecules, induce intense oxidative stress, upset the intracellular pH (pHi) homeostasis, and initiate a significant amount of necrosis. Experimental research in vivo has shown that the lead photooxidant inhibits tumor growth, suppresses metastasis, and activates the immune system in a manner that is effective. In this study, metal-enhanced photooxidation is validated along with its subsequent chemotherapeutic applications, providing evidence for the development of localized photooxidants that will directly damage intracellular biomolecules and decrease pHi as a strategy for the development of effective treatments.

There are a number of platinum compounds that are commonly used in the clinic as anticancer drugs[67]. To combat drug resistance and reduce side effects, new generations of metal-based anticancer agents are urgently needed[68]. It is known that photoactivated metal complexes can provide both temporal and spatial control over drug activation, and they have shown remarkable potential for cancer therapy[69]. Several tin-, lutetium-, palladium-, and ruthenium-based photosensitizers are currently being tested in clinical trials or are approved for use in cancer PDT[70]. In the fight against drug resistance, PDT has the advantage of high therapeutic efficacy at low doses, as it is capable of catalyzing a wide range of chemical reactions[71]. In spite of this, clinical resistance to photosensitizers has been reported in a few instances[72].

A number of mechanisms contribute to the development of PDT resistance, including hypoxia, detoxification by intracellular antioxidant systems, the induction of stress response genes, and the efflux of drugs by P-glycoproteins. It is evident from these considerations that new generations of photosensitizers with incredible MOA (mechanisms of action) for the treatment of cancer are urgently needed. PDT's MOA is heavily dependent on oxygen to function, which results in limiting the application of the PDT to cancer cells since they grow mainly in a hypoxic environment[73]. Researchers have investigated a number of novel photosensitizers with new mechanism of action against various tumors in several studies. For example, there have been studies done targeting photoredox catalysis to treat cancer cells[74].

According to this study, photoredox catalysis may be able to give an oxygen-independent MOA for combating this issue[74]. The aim of the

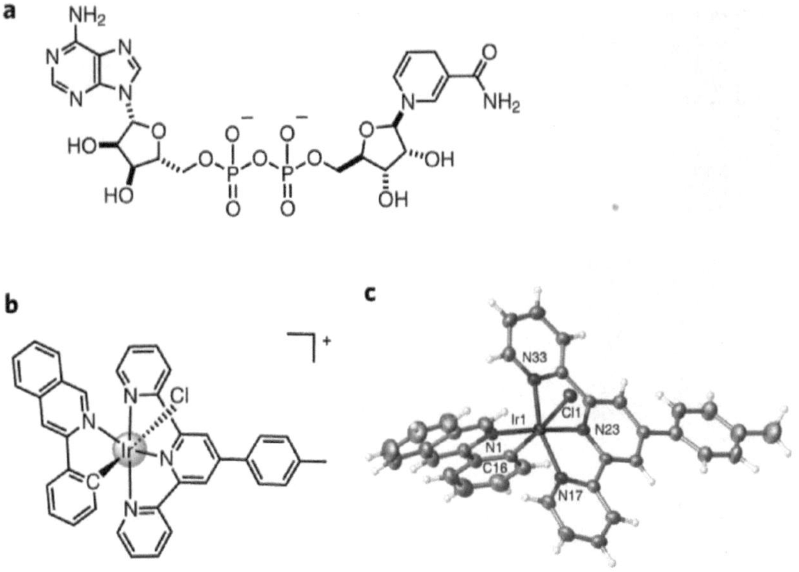

FIGURE 4.7 Structures of compounds. (a,b) Line structures of NADH (a) and complex 1 (b). (c) The X-ray crystal structure of complex 1. (Adapted with permission from ref. [74].)

research is to make an Ir(iii) photocatalyst that is highly oxidative, and is very phototoxic towards both hypoxic and normoxic tumor cells. Complex 1 (**Figure 4.7**) photocatalytically oxidizes NADH (1,4-dihydronicotinamide adenine dinucleotide), generating NAD• radicals in biological media, with a high turnover rate. Additionally, NADH and complex 1 together photoreduce the cytochrome c protein under hypoxia in a synergistic manner. Based on the DFT (density functional theory) calculations, it has been demonstrated that there is π stacking in the adducts of NADH and complex 1, enabling the transfer of single electrons under photoinduced conditions (**Figure 4.8**). Complex 1 localizes in the mitochondria of cancer cells and interferes with electron transport in the cells by means of NADH photocatalysis. When light is irradiated onto complex 1, it causes the depletion of NADH, an intracellular redox imbalance, and an immunogenic apoptotic cell death in cancer cells (**Figure 4.9**). A new approach to cancer phototherapy is offered by using this photocatalytic redox imbalance.

As the primary medical remedy for cancer, the most effective treatment is chemotherapy. The efficacy of chemotherapy treatment is limited, however, by the presence of resistance to anticancer drugs, which is the hallmark of

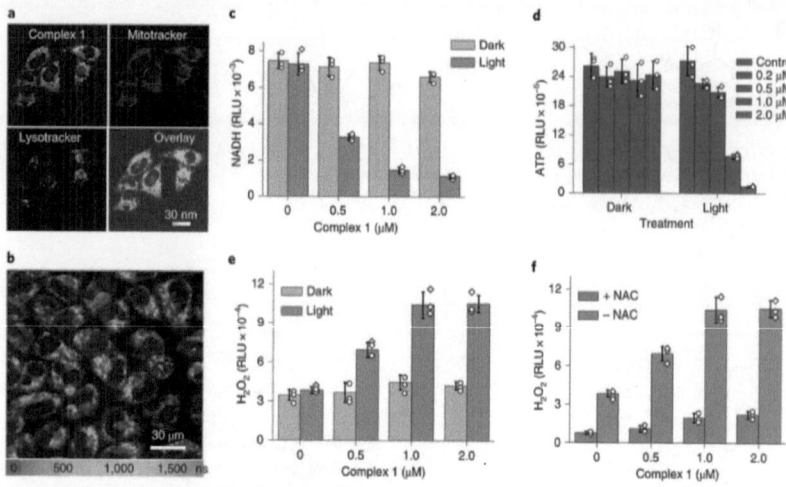

FIGURE 4.8 Cellular localization and cellular response after irradiation. (Adapted with permission from ref. [74].)

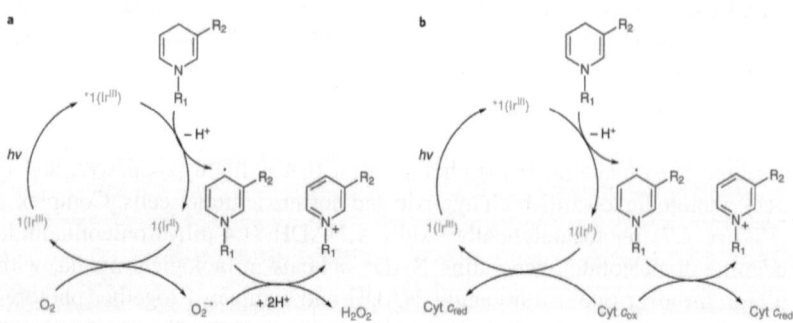

FIGURE 4.9 Photocatalytic cycle for oxidation of NADH by complex 1, showing the production of NAD radicals, involvement of oxygen, and reduction of Cyt c. (Adapted with permission from ref. [74].)

malignant tumors. Because anticancer drugs are not cancer selective, toxic, and have dose complications, the drug design is less attractive as a long-term cancer treatment. Furthermore, it is important to recognize that most of the anticancer drugs are also restricted to passive diffusion within the body, making them unsuitable to target the site of action inside the body. In addition, it is important to note that CSCs, also known as tumor-initiating cells, are believed to play a significant role in the development of drug resistance and relapse of cancer since they are capable of performing self-renewal and promoting metastasis[75]. As a matter of fact, there are almost no effective procedures to remove CSCs apart from the use of molecular inhibitors of cancer stemness[76]. In spite of this, these strategies do not yield enough efficacy. It remains a major challenge for scientists to develop strategies for inhibiting CSCs.

Using nanomaterials in conjunction with physical procedures, like optical excitation, acoustic excitation, and magnetic excitation, it is possible to control the activities of cellular with a temporal and spatial selectivity[77]. These approaches, however, are ineffective because they were developed independently of molecular genetics and do not target specific cellular functions. In order to achieve these goals, optogenetics could be used to selectively activate or inhibit neurons utilizing genetically encoded light-sensitive ion channels or opsins. In spite of this, Due to the limited penetration of visible light into tissues, its clinical applications have been limited[78].

There is an urgent need for strategies to eradicate CSCs, since CSCs are resistant to anticancer drugs and are a leading cause of treatment failures, relapses, and metastasis. An infrared-light-activatable nanocomplex that inhibited the growth of CSCs through photothermogenetic inhibition has been reported[79]. The study revealed that photoactive functional nano-carbon complexes possess unique properties, including high dispersibility in water, homogeneous particle morphology, excellent photothermal stability, rapid photoresponsivity, and powerful photothermal conversion. In addition to this, the present biologically permeable NIR-II (second NIR) light-induced nanocomplexes are capable of photothermally triggering calcium influx into target cells that are overexpressing a TRPV2 (transient receptor potential vanilloid family type 2, **Figure 4.10**). As a result of the combination of genetic engineering and nanomaterial design, cancer cells are effectinely eliminated and the stemness of cancer cells is suppressed both in vivo and in vitro (**Figures 4.11** and **4.12**).

Last but not least, a molecular analysis of the mechanisms involved for inhibiting cancer stemness shows that dysregulation of the Wnt/beta-catenin signaling pathway mediated by calcium is responsible for inhibiting cancer stemness. As a result of the technological concept, there may be innovative treatments in the future to address the global threat of tumors that are refractory.

Due to the excellent biosafety of hydrogen therapy, it has been gaining a lot of attention as a new emerging treatment strategy for a variety of diseases. Hydrogen, on the other hand, is difficult to accumulate in local lesions due to its low solubility and high diffusivity. The development of na NIR-driven water splitting nanoplatform for the production of H_2 for using in cascade-amplification synergetic cancer therapy has been reported in the literature[80]. CSNPs (core–shell nanoparticles) of $NaGdF_4:Yb,Tm/g-C_3N_4/Cu_3P$ (UCC)

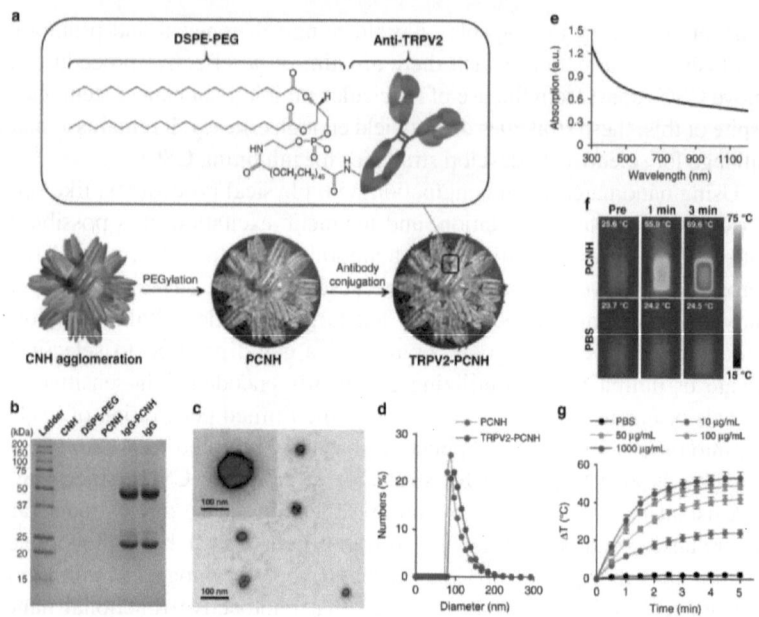

FIGURE 4.10 Syntheses and characterization TRPV2-PCNH. (Adapted with permission from ref. [79].)

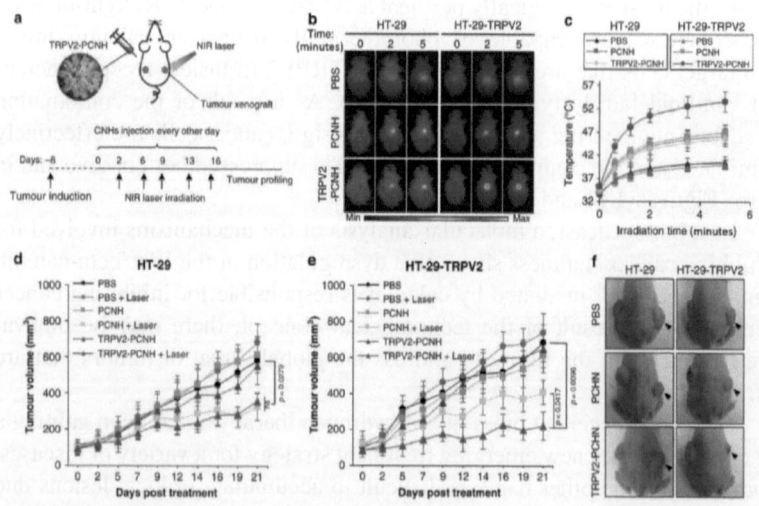

FIGURE 4.11 Laser-induced TRPV2-PCNH inhibits tumor progression in in vivo xenograft models with TRPV2 overexpression. (Adapted with permission from ref. [79].)

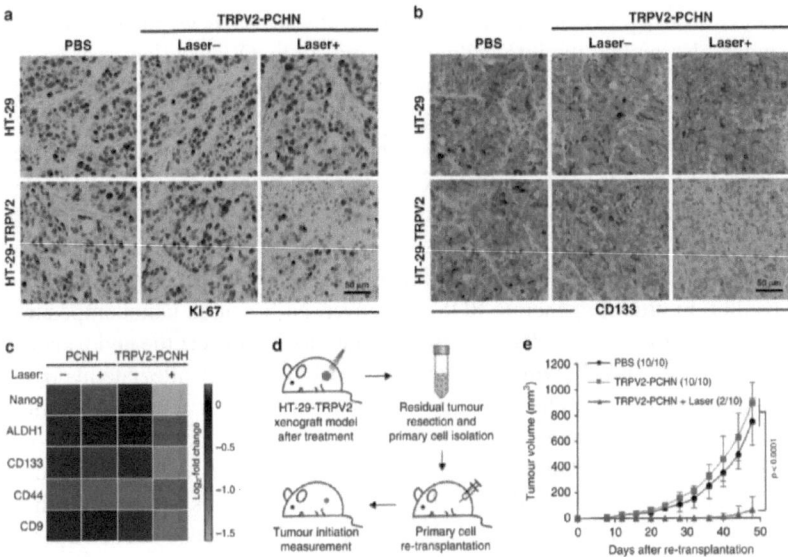

FIGURE 4.12 Laser-induced TRPV2-PCNH inhibits tumor re-initiation in in vivo xenograft models with TRPV2 overexpression. (Adapted with permission from ref. [79].)

nanocomposites as core encapsulated with ZIF-8 (zeolitic imidazolate framework-8) modified with folic acid as shell are designed and synthesized in this. Because of the acid-responsive ZIF-8 shell, the enhanced retention effect and permeability, and the folate receptor-mediated endocytosis, CSNPs are selectively captured by tumor cells.

As a result of laser irradiation at 980 nm, CSNPs exhibit a high production capacity of ROS and H_2, in addition to having an appropriate photothermal conversion temperature after being exposed to laser light. Furthermore, as the temperature rises, so does the Fenton reaction rate of Cu(I) with H_2O_2 and as a result, the curative effect of CDT (chemodynamic therapy). An excess of GSH (glutathione) in the TME can deplete positive holes generated in the valence band of the g-C_3N_4 in the g-C_3N_4/Cu_3P heterojunction. It has also been shown that Cu(II) was reduced to Cu(I) by GSH, which ensures a continuous Fenton reaction to take place. Consequently, H_2-mediated cascade-amplifying multimodal synergetic therapy is enabled through a nanoplatform based on NIR-based H_2 production.

The hypoxia in the TME has been found to inhibit the efficacy of PDT. NIR-driven intracellular photocatalytic O_2 evolution on Z-Scheme Ni_3S_2/$Cu_{1.8}S$@HA for hypoxic tumor therapy was reported[81]. Using Ni_3S_2/$Cu_{1.8}S$

nanoheterostructures, it was possible to synthesize a new photosensitizer that can also realize the intracellular photocatalytic O_2 evolution in order to relieve hypoxia in the TME and enhance the PDT at the same time. Due to the narrow band gap (below 1.5 eV), the NIR (808 nm) can be used to stimulate the separation of electrons and holes. A novel Z-scheme nanoheterostructure possessing a higher redox ability is demonstrated by both experimental data and DFT calculations, and as a result the photoexited holes have sufficient potential to oxide H_2O directly into oxygen.

In the meantime, the photostimulated electrons can capture the dissolved O_2 in order to generate ROS. Moreover, $Ni_3S_2/Cu_{1.8}S$ nanocomposites also possess the catalase-/peroxidase-like activity to convert the endogenous H_2O_2 into $\cdot OH$ and O_2, which not only cause CDT but also alleviate hypoxia to assist the PDT as well. They also have a high NIR harvesting capability and a high photothermal conversion efficiency as a result of their narrow band gap. It is noted that the nanocomposites also exhibit novel biodegradation and can be metabolized as well as eliminated via feces and urine within a two-week period. As a result of the present single electrons in Ni/Cu ions, $Ni_3S_2/Cu_{1.8}S$ exhibits the magnetic resonance imaging capability (MRI). To make sure that the cancer cells were specifically targeted, HA (hyaluronic acid) was grafted outside and $Ni_3S_2/Cu_{1.8}S@HA$ integrated PDT, CDT, and PTT to exhibit the great anticancer efficiency for hypoxic tumor elimination.

Due to its minimal invasiveness, high efficacy, and low side effects, phototherapy, particularly PDT and PTT, has emerged as a promising therapeutic technique to treat skin cancers. The clinical effectiveness of single-modality therapy in treating skin cancer, whether it is PTT or PDT, is, however, limited. It has been frequently reported that PTT and PDT can be used together in various applications. The difference between PTT and PDT is that each requires its own photoagents and sources of excitation light, resulting in significant challenges when it comes to clinical transformation. The combination of PTT and PDT for cutaneous squamous cell carcinoma by gold nanoclusters under a single NIR laser irradiation has been reported to be effective[82]. This study aimed to investigate the utilization of biocompatible gold nanoclusters $Au_{25}(Capt)_{18}$ for the concurrent PDT and PTT treatments of cSCC (cutaneous squamous cell carcinoma) using an NIR laser at an 808 nm wavelength.

It is demonstrated that $Au_{25}(Capt)_{18}$ nanoclusters were capable of significantly suppressing proliferation of cSCC XL50 cells in vitro as well as inhibiting tumor growth in SKH-1 mice by utilizing their strong photothermal stability, ability to generate singlet oxygen, and high light-thermal conversion efficiency. PTT and PDT, both of which are known to kill tumor cells at 808 nm, were estimated to have 28.86% and 71.14% of the tumor-cell-killing ability, respectively, by using an ROS scavenger to quench the effect of PDT on tumor cells. It was observed after one course of concurrent PDT and PTT that tumor-infiltrating CD4+ T and CD8+ T cells were detected. It was found that the $Au_{25}(Capt)_{18}$ nanoclusters have low adverse effects based on preliminary toxicity studies. Using a simple nanostructure, the study reported that it was possible to use PDT and PTT simultaneously to kill cSCC and to induce an anti-tumor immune response simultaneously. Researchers may be able to develop effective photoagents for future, synergistic applications of different phototherapies with targeted immunological responses for the treatment of cancer through this study.

It appears that using NIR light for PDT could be a promising method to circumvent the limitations of the current PDT in which visible light has a limited penetration depth into the tissues. There have been reports of NIR-II light activated PDT with proteins capped on gold nanoclusters[83]. The study employed alkyl thiolated AuNCs (gold nanoclusters) that were co-modified with CAT (catalase) and HSA (human serum albumin), in order to create a multifunctional, optical, and theranostic nano-agent. It has been shown that the AuNC@HSA/CAT system is capable of producing singlet oxygen under excitation by a 1064-nm laser. This laser lies within the NIR-II (second NIR window) and features much lower scattering and tissue absorption, thus enabling NIR-II-triggered PDT.

Using an HSA coating on the NPs greatly improved their physiological stability, which has shown to have effective tumor retention after intravenous injection, as confirmed by the detection of AuNC fluorescence. The presence of CAT in the NPs also resulted in the decomposition of tumor endogenous H_2O_2 to generate oxygen, thereby enhancing the effectiveness of PDT by relieving tumor hypoxia. There is a remarkable increase in tissue penetration when NIR-II-triggered PDT is compared with conventional PDT using visible light. Accordingly, the study created a new type of photosensitizing nano-agent that simultaneously enables in vivo fluorescence imaging, tumor hypoxia relief, and NIR-II light-induced in vivo PDT for the treatment of cancer.

There is a need for a deeper understanding of the structure-activity relationship in the development of organic supramolecular photocatalytic materials. The Supramolecular packing dominant photocatalytic oxidation performance of PDI has been reported and was shown to affect anticancer properties[84]. The effects of stacking of H/J types on the photocatalytic mechanism of PDI and its activity have been investigated. As a face-to-face arrangement, H-aggregates have higher π-electron conjugation and show more semiconductor characteristics, which results in higher carrier separation and migration efficiency under irradiation. Whereas, due to its low π-electron conjugation caused by head-to-tail stacking mode, J-aggregates exhibit more molecular properties.

As a result, H-aggregated PDI mainly forms superoxide radicals (O^{2-}) and holes (h^+) through ET (electron-transfer). In contrast, J-aggregated PDI mainly generates singlet oxygen species ($^1O_2$) by means of an EnT process (energy transfer). Benefiting from the stronger oxidizability of O^{2-} and h^+, H-aggregated PDI shows higher photocatalytic activity for small molecule degradation and oxygen evolution under visible light. It has also been found that J-aggregated PDI exhibits potential uses in photocatalytic anti-cancer treatment due to its high $^1O_2$ quantum yields when exposed to red light. This research may provide guidance for the development of supramolecular organic photocatalytic materials in the future.

In cancer therapy, the rapid, complete, targeted, and safe treatment of tumors remains one of the most important issues to address. Using a supramolecular porphyrin photocatalyst, a rapid elimination of solid tumors was observed by generating holes in the photogenerated system[85]. With the irradiation of 600 to 700 nm wavelength, it is possible to treat solid tumors by a supramolecular photocatalyst, Nano-SA-TCPP (**Figure 4.13**). In just 10 minutes, solid tumors with a volume of 100 mm^3 can be eliminated from the body (**Figures 4.14** and **4.15**). A 50-day survival rate for mice treated

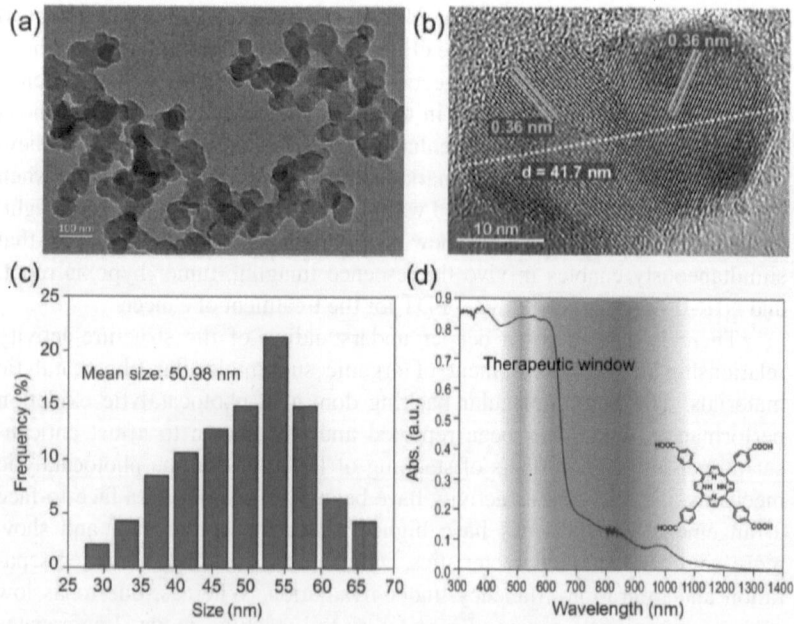

FIGURE 4.13 The characterization of the Nano-SA-TCPP. (Adapted with permission from ref. [85].)

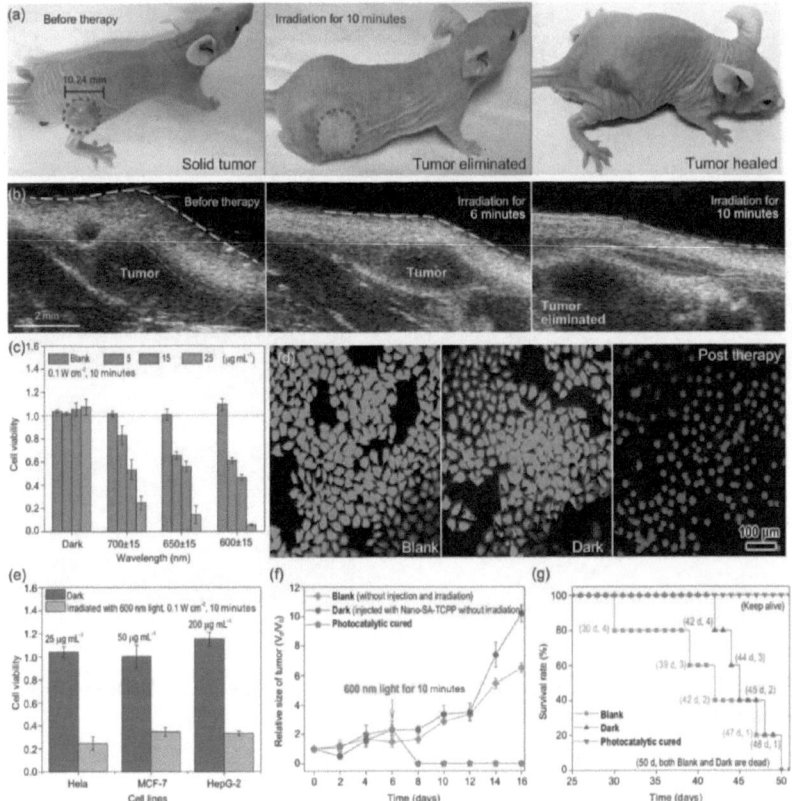

FIGURE 4.14   The photocatalytic cancer therapy with Nano-SA-TCPP. (Adapted with permission from ref. [85].)

with the photocatalytic therapy was increased from zero to 100% after the treatment was applied.

A breakthrough was made due to the rupture of the cell membrane and the loss of cytoplasm caused by photogenerated holes inside cancer cells. It has been found that porphyrin-based photocatalysts can be internalized in a targeted manner by cancer cells as a result of the size selection effect, without entering the normal cells in the process. There is no toxicity or side effects associated with this therapy for normal cells or organisms. Further, the photocatalytic therapy has been found to be effective on a variety of cancer cell lines as well. As a result of its high efficiency, safety, and universality, photocatalytic therapy can be considered as a new lancet that can be used to conquer cancer.

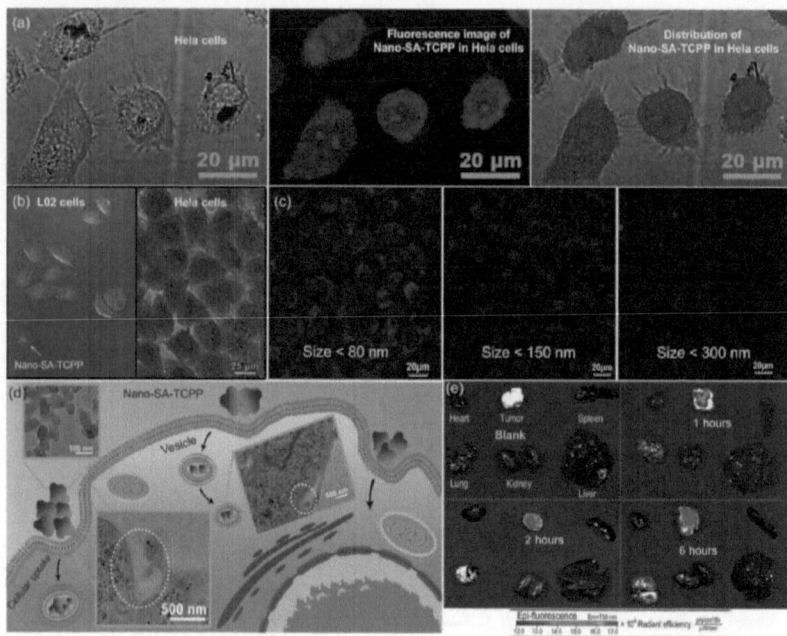

FIGURE 4.15 The cellular uptake and targeted property of Nano-SA-TCPP. (Adapted with permission from ref. [85].)

4.6 CANCER THERAPY USING OPTICALLY-CONTROLLED BACTERIAL METABOLITES

As a consequence of the exceptional tumor colonizing ability of some bacteria, bacteria-mediated treatments have attracted great attention, particularly for the treatment of cancer[86]. Aside from their ability to target tumors, chemical studies have also been conducted to find ways in which drug-loaded nanomaterials can be directly conjugated to tumor-targeting bacteria such as Magnetococcus and Salmonella[87]. It has also been demonstrated that genetic engineering can also be used as a biological strategy in the process of genetically modifying bacteria for the production of in situ anti-cancer agents[88]. There are, however, some inherent drawbacks to both of these strategies. The loss of transgene expression in biological approaches and the limited carrying capacity of chemical approaches have resulted in insufficient dosages of anti-cancer agents and unsatisfactory therapeutic results[89]. In spite of this, neither synthetic approaches nor genetic approaches alone can fully satisfy the current requirements in terms of both high efficiency and stability at present.

Several bacteria have shown the ability to spontaneously convert non-toxic compounds into antineoplastic products[90]. It should be noted, however, that these natural bioreactions are usually too weak to be able to achieve satisfactory therapeutic results. Some bacteria possess the capacity of driving intracellular reactions at the expense of exogenous electrons. The idea was derived from this phenomenon that charging bacteria with abundant exogenous electrons would lead to the control of their intrinsic metabolic activities as well as the enhancement of their latent anticancer properties. As of recent, a large number of nanoscale photocatalytic materials such as C_3N_4 and CdS have attracted considerable interest due to their ability to continuously convert light into electrical energy through their photoelectric converting ability[91,92]. In addition to the fact that these kinds of materials are able to "charge" bacteria in order to strengthen their metabolic activities, but they also go beyond loading limits in order to produce an abundance of anti-cancer agents. It is possible that this universal strategy could create a biotic/abiotic loop that would allow the synthesis of NO from NO3- under the influence of light.

Optimally controlled bacterial metabolites have been demonstrated to be effective in the treatment of cancer[93]. A photocatalytic system was fused with a tumor-targeted bacterium, so as to achieve a biotic/abiotic hybrid for the generation of NO by light control. The *E. coli* MG1655, a non-pathogenic bacterium that has both nitrate/nitrite reductase expression as well as tumor targeting, was selected for modification to support the achievement of this goal[94]. It has been demonstrated that CCN (carbon dot doped carbon nitride) with suppressed free radical generation capability can be used for in situ photoelectric conversion[95]. Additionally, *E. coli* and CCN are assembled to form CCN@E. coli through the electrostatic interaction.

As part of the study, the concept of photo-controlled bacterial metabolite therapy (PMT) has been proposed, which utilizes modified CCN@E. coli to metabolize NO^{3-} in order to produce antineoplastic NO under photo-irradiation for use in cancer treatment (**Figure 4.16**). Moreover, it is also possible to verify its detailed mechanism by utilizing the isotope labeling method, as well as a proteomics study. As far as mammals are concerned, endogenic NO is produced by nitric oxide synthase by enzymatic means from L-arginine. As a result of the physiological processes that NO goes through, it can spontaneously be converted into NO^{3-}, which can have no effect on mammalian cells. There is no doubt that the PMT system is dependent on NO^{3-} as the primary source of nitrogen[96]. Consequently, the irreversible generation of NO is converted into a circular reaction that maximizes the bioavailability of NO

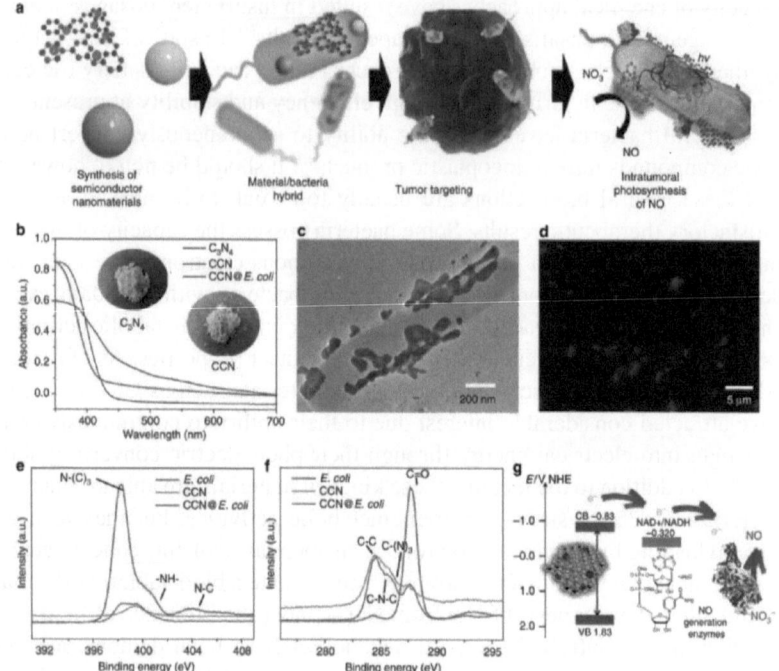

FIGURE 4.16 Characterization of PMT system. (a) Schematic diagram of the preparation of PMT system. (b) UV–Vis absorption spectra of as-prepared CCN. (c) TEM image of CCN@E. coli. (d) Spinning disk confocal microscope image of CCN@E. coli. (e) XPS de-convoluted spectra for the N1s orbitals of *Escherichia coli*, CCN, and CCN@E. coli. (f) XPS de-convoluted spectra for the C1s orbitals of *E. coli*, CCN, and CCN@E. coli. (g) Schematic illustration for the photoelectron transport among CCN, electron acceptor, and NO generation enzymes. (Adapted with permission from ref. [93].)

in the body. It is clear that PMT is one of the most significant advancements in the area of bacterial cancer treatment since it optimizes the biomedical applications of biotic/abiotic hybrid systems.

Under light irradiation, the photoelectrons produced by C_3N_4 can be transferred to *E. coli*, which promotes the enzymatic reduction of endogenous NO_3 to cytotoxic NO by a 37-fold increase (**Figures 4.17** and **4.18**). It has been demonstrated that C_3N_4 loaded bacteria are perfectly accumulated throughout the tumor in a mouse model, and the PMT treatment results in an estimated 80% inhibition of the growth of the tumor. In this way, synthetic materials remodeled microorganisms with the aim of regulating focal microenvironments and increasing therapeutic effectiveness may be used.

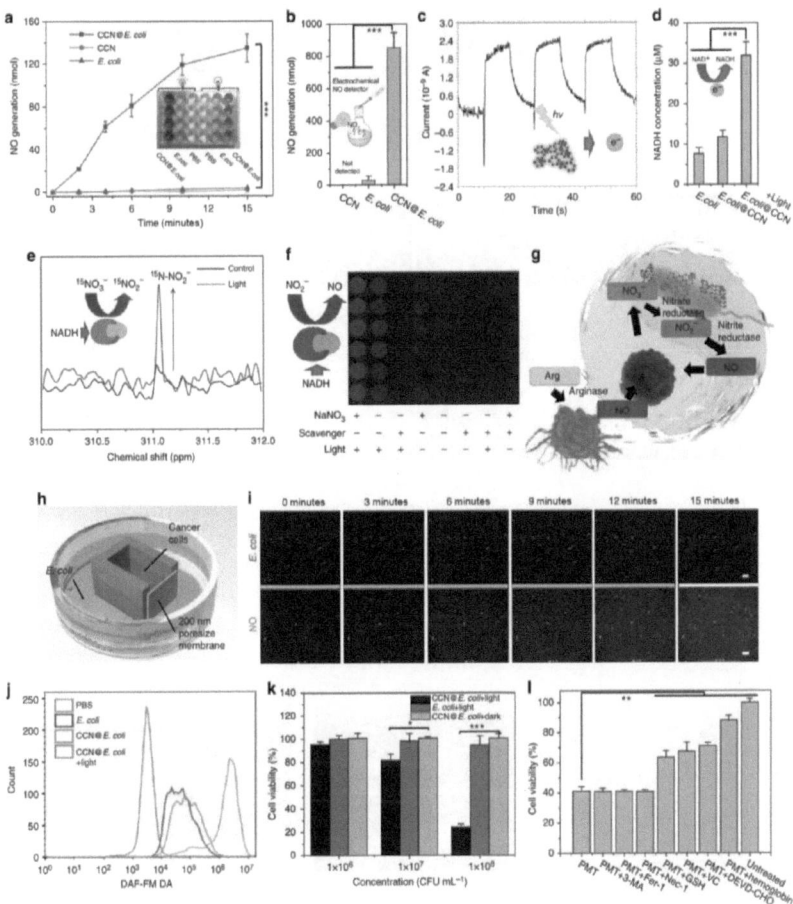

FIGURE 4.17 In vitro study of PMT system. (a) Griess method for quantitative determining the NO generation of CCN@E. coli. (b) Electrochemical method for monitoring the cumulative NO in the gas phase produced by CCN@E. coli. (c) Transient photocurrent responses of CCN. (d) Intracellular NADH level of wide-type E. coli. (e) 15N-NMR for monitoring the in situ CCN@E. coli NO₃ − metabolism with or without light irradiation. (f) Luminol chemiluminiscence assay for qualitative determining NO generation of CCN@E. coli, E. coli, and CCN. (g) Schematic illustration for the optically controlled NO metabolism close loop. (h) Schematic diagram of the 3D-printing co-culture system. (i) Images from the co-culture system time series sequentially observing CCN@E. coli movement and NO generation. (j) Flow cytometry for measuring the intracellular NO concentration of 4T1 cells after various treatment. (k) Cell viability of 4T1 cells co-cultured with CCN@E. coli. (l) Cell viability of 4T1 cells co-cultured with CCN@E. coli. (Adapted with permission from ref. 93.)

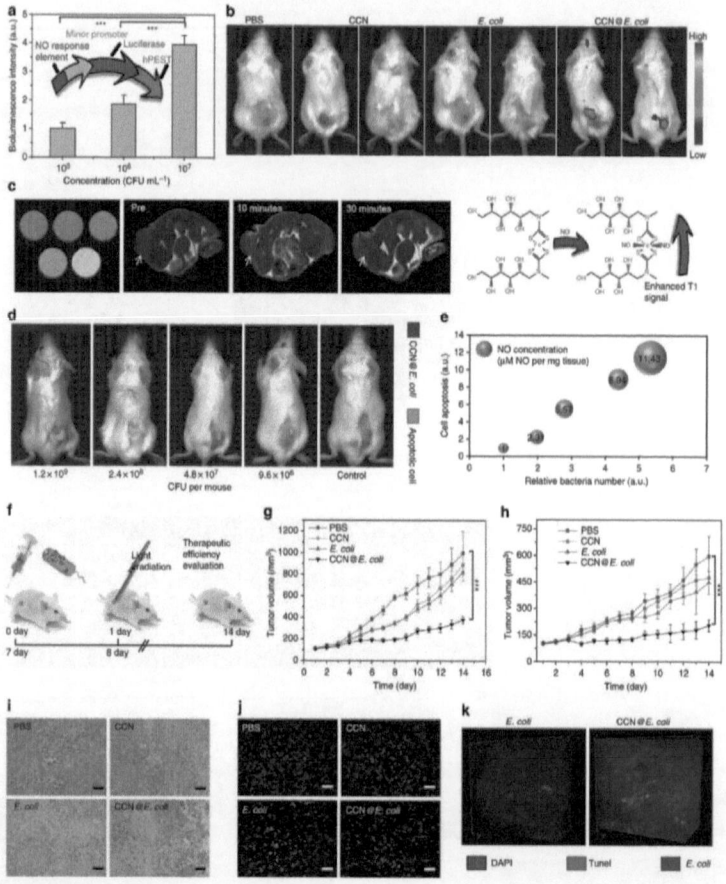

FIGURE 4.18 In vivo NO generation and anti-cancer effect of PMT system. (Adapted with permission from ref. [93].)

4.7 CANCER THERAPY USING PHOTOACTIVATABLE METABOLIC WARHEADS

In recent years, photoactivatable molecules have been developed that enable the ablation of malignant cells in response to light. However, current agents can prove ineffective at early stages of disease when the target cells are similar to healthy surrounding tissues. It has been reported that photoactivatable metabolic warheads can be used to precisely and safely destroy target cells in vivo[97]. In this work, a chemical platform based on amino-substituted benzoselenadiazoles is described in order to build photoactivatable probes that mimic native metabolites to be used as indicators of onset and progression of diseases. In a series of synthetic derivatives of benzoselenadiazole scaffolds, it has been possible to identify the key chemical groups in the scaffold that are responsible for the photodynamic activity of the scaffold (**Figure 4.19**).

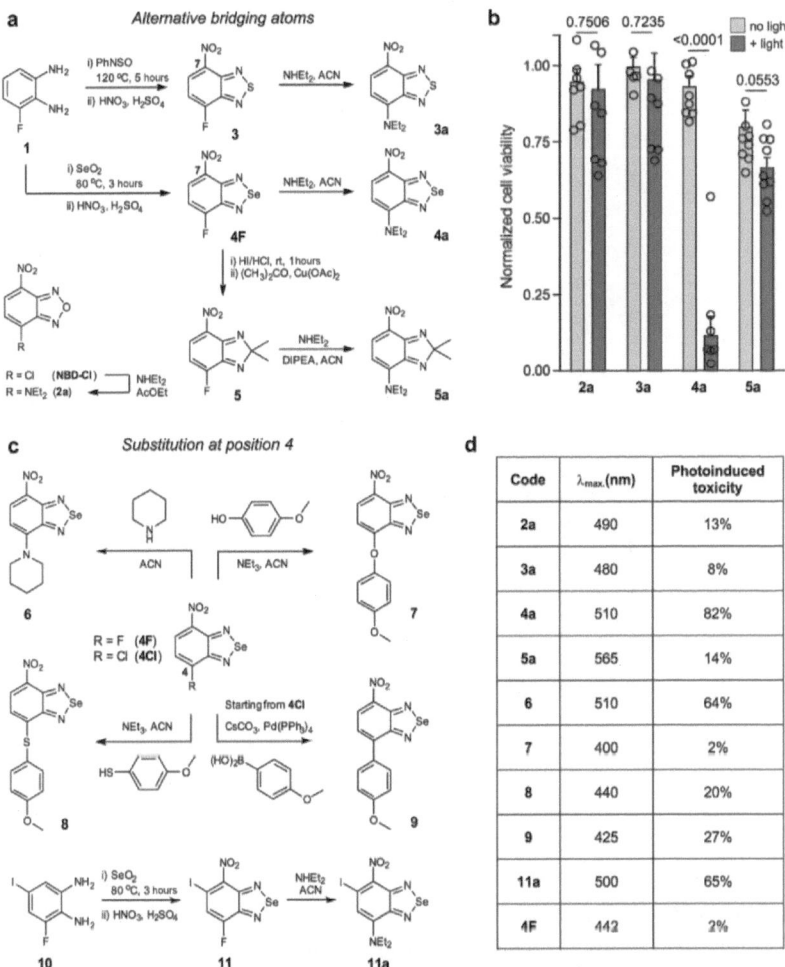

FIGURE 4.19 Synthetic routes for the preparation of small photosensitizers. (Adapted with permission from ref. [97].)

This work was followed by the development of photosensitive metabolic warheads, which could be used to target cells associated with various types of diseases, including bacterial infections and cancers. It has been demonstrated in the study that versatile benzoselenadiazole metabolites have the ability to selectively kill pathogenic cells, but not healthy cells, after exposure to non-toxic visible light as well as to reduce any potential adverse effects in vivo (**Figures 4.20** and **4.21**). It gives powerful tools for exploiting cellular metabolic signatures as a means of developing safer therapeutics and surgical approaches based on this chemical platform.

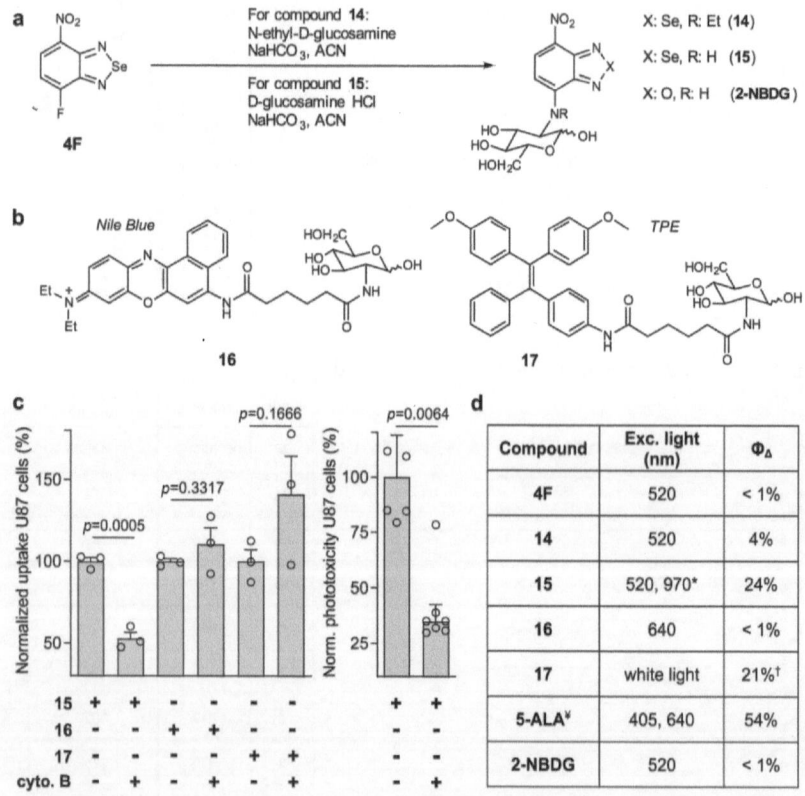

FIGURE 4.20   D-Glucose derivatives of benzoselenadiazole – but not of other photosensitizers – are recognized by GLUT transporters. (Adapted with permission from ref. [97].)

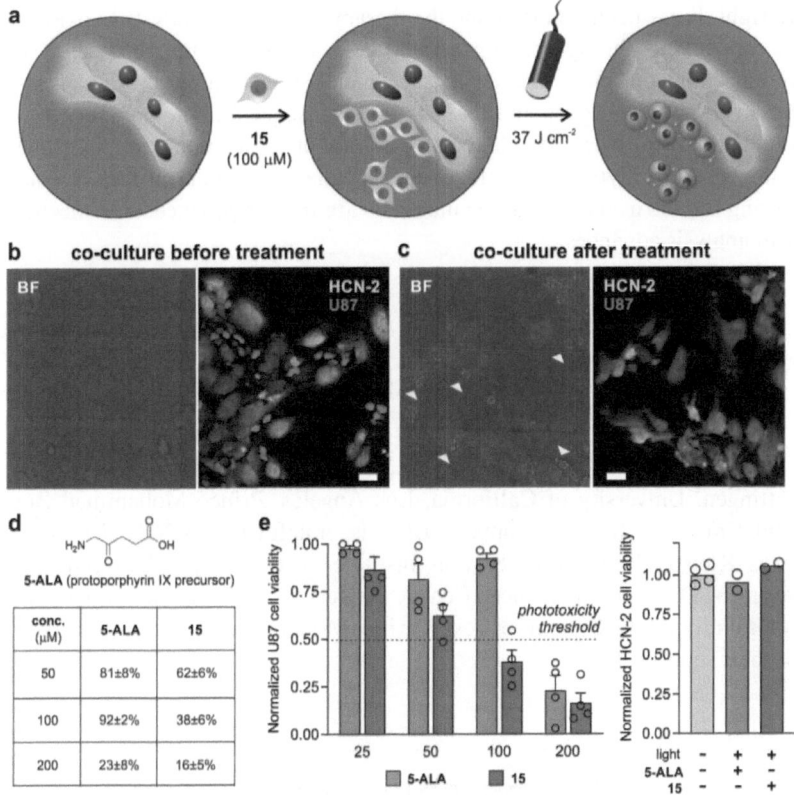

FIGURE 4.21 Compound 15 selectively ablates human glioblastoma cancer cells under clinical PDT conditions and in the presence of non-cancerous human brain cells. (Adapted with permission from ref. [97].)

4.8 CONCLUSIONS

This chapter presents a review of some of the current progresses in the field of photo-induced drug-free sustainable therapeutic approaches for cancer treatment. In the development of this novel cancer treatment, different photocatalysts were used as the catalysts. It has been shown that solid tumors can be rapidly eliminated under light with proper photocatalysts if the conditions are right. It has been reported that the therapy uses photocatalysts to oxidize the cancer cells through photogenerated holes created by the photocatalyst. By enriching the photocatalyst in cancer cells and tumors, no damage is caused to normal organs and tissues. As a consequence, the organism can completely metabolize the photocatalyst without bioaccumulating it. In the future, photocatalytic cancer treatment will play an important role in conquering tumors as photocatalytic materials are further enhanced and mechanism analysis advances.

ACKNOWLEDGMENTS

AD is grateful to CEA-Grenoble, Joseph Fourier University, University of Göttingen, University of California, Los Angeles, Prince Mohammad Bin Fahd University for their support. BKB is grateful to US NIH, US NCI, Texas Kleberg Foundation, Stevens Institute of Technology, University of Texas M. D. Anderson Cancer Center, University of Texas-Pan American, Community Health Systems of Texas, Prince Mohammad Bin Fahd University for their support.

REFERENCES

1. Torre LA, Bray F, Siegel RL, Ferlay J, Lortet-Tieulent J, Jemal A. Global cancer statistics, 2012. *CA Cancer J Clin.* 2015;65(2):87–108. doi:10.3322/caac.21262
2. Ahles TA, Saykin AJ. Candidate mechanisms for chemotherapy-induced cognitive changes. *Nat Rev Cancer.* 2007;7(3):192–201. doi:10.1038/nrc2073
3. Naredi P, La Quaglia MP. The future of trials in surgical oncology. *Nat Rev Clin Oncol.* 2015;12(7):425–431. doi:10.1038/nrclinonc.2015.72
4. Sharifi S, Behzadi S, Laurent S, Forrest ML, Stroeve P, Mahmoudi M. Toxicity of nanomaterials. *Chem Soc Rev.* 2012;41(6):2323–2343. doi:10.1039/c1cs15188f
5. Bao Z, Li K, Hou P, Xiao R, Yuan Y, Sun Z. Nanoscale metal–organic framework composites for phototherapy and synergistic therapy of cancer. *Mater Chem Front.* 2021;5(4):1632–1654. doi:10.1039/D0QM00786B
6. Bown SG. Phototherapy of tumors. *World J Surg.* 1983;7(6):700–709. doi: 10.1007/BF01655209
7. Das A. LED light sources in organic synthesis: an entry to a novel approach. *Lett Org Chem.* 2022;19(4):283–292.
8. Das A. Recent developments in semipolar InGaN laser diodes. *Semiconductors.* 2021;55(2):272–282. doi:10.1134/S106378262102010X

9. Das A. A systematic exploration of InGaN/GaN quantum well-based light emitting diodes on semipolar orientations. *Opt Spectrosc.* 2022;130(3):137–149. doi:10.1134/S0030400X2203002X

10. Rozhkova EA, Ulasov I, Lai B, Dimitrijevic NM, Lesniak MS, Rajh T. A high-performance nanobio photocatalyst for targeted brain cancer therapy. *Nano Lett.* 2009;9(9):3337–3342. doi:10.1021/nl901610f

11. Das A, Banik BK. Chapter 13 - Synthesis of Natural Products by Photochemistry. In: Banik BK, ed. *Green Approaches in Medicinal Chemistry for Sustainable Drug Design (Second Edition).* Vol 2. Advances Green and Sustainable Chemistry. Elsevier; 2024:259–283. doi:10.1016/B978-0-443-16164-3.00013-3

12. Ferrari M. Cancer nanotechnology: opportunities and challenges. *Nat Rev Cancer.* 2005;5(3):161–171. doi:10.1038/nrc1566

13. Das A, Ashraf MW, Banik BK. Thione derivatives as medicinally important compounds. *ChemistrySelect.* 2021;6(34):9069–9100. doi:10.1002/slct.202102398

14. Das A, Banik BK. Advances in heterocycles as DNA intercalating cancer drugs. *Phys Sci Rev.* 2023;8(9):2473–2521. doi:10.1515/psr-2021-0065

15. Das A, Banik BK. 4 Advances in Heterocycles as DNA Intercalating Cancer Drugs. In: Krishna Banik B, Banerjee B, eds. *Heterocyclic Anticancer Agents.* De Gruyter; 2022:111–160. doi:10.1515/9783110735772-004

16. Banik BK, Das A. *Natural Products as Anticancer Agents.* Elsevier; 2023.

17. Das A, Banik BK. Chapter 1 - Anticancer Activity of Natural Compounds from Leaves of the Plants. In: Krishna Banik B, Das A, eds. *Natural Products as Anticancer Agents.* Elsevier; 2024:3–48. doi:10.1016/B978-0-323-99710-2.00008-1

18. Das A, Banik BK. Chapter 2 - Anticancer Activity of Natural Compounds from stems/barks of the Plants. In: Krishna Banik B, Das A, eds. *Natural Products as Anticancer Agents.* Elsevier; 2024:49–86. doi:10.1016/B978-0-323-99710-2.00010-X

19. Das A, Banik BK. Chapter 3 - Anticancer Activity of Natural Compounds from Roots of the Plants. In: Krishna Banik B, Das A, eds. *Natural Products as Anticancer Agents.* Elsevier; 2024:87–132. doi:10.1016/B978-0-323-99710-2.00009-3

20. Das A, Banik BK. Chapter 4 - Anticancer Activity of Natural Compounds from Fruits and Vegetables. In: Krishna Banik B, Das A, eds. *Natural Products as Anticancer Agents.* Elsevier; 2024:133–178. doi:10.1016/B978-0-323-99710-2.00001-9

21. Banik BK, Das A. Chapter 5 - Anticancer Activity of Natural Compounds from Marine Animals. In: Krishna Banik B, Das A, eds. *Natural Products as Anticancer Agents.* Elsevier; 2024:181–236. doi:10.1016/B978-0-323-99710-2.00012-3

22. Banik BK, Das A. Chapter 6 - Anticancer Activity of Natural Compounds from Marine Plants. In: Krishna Banik B, Das A, eds. *Natural Products as Anticancer Agents.* Elsevier; 2024:237–284. doi:10.1016/B978-0-323-99710-2.00003-2

23. Das A, Banik BK. Chapter 7 - Anticancer Activity of Natural Compounds from Bacteria. In: Krishna Banik B, Das A, eds. *Natural Products as Anticancer Agents.* Elsevier; 2024:287–328. doi:10.1016/B978-0-323-99710-2.00011-1

24. Banik BK, Das A. Chapter 8 - Anticancer Activity of Natural Compounds from Fungi. In: Krishna Banik B, Das A, eds. *Natural Products as Anticancer Agents*. Elsevier; 2024:329–366. doi:10.1016/B978-0-323-99710-2.00004-4

25. Banik BK, Das A. Chapter 9 - Anticancer Drugs from Hormones and Vitamins. In: Krishna Banik B, Das A, eds. *Natural Products as Anticancer Agents*. Elsevier; 2024:369–414. doi:10.1016/B978-0-323-99710-2.00006-8

26. Banik BK, Das A. Chapter 10 - Future Prospects in Anticancer Natural Products. In: Krishna Banik B, Das A, eds. *Natural Products as Anticancer Agents*. Elsevier; 2024:415–426. doi:10.1016/B978-0-323-99710-2.00002-0

27. Das A, Banik BK. Combatting the coronavirus utilizing natural cinnamon and its derived products. *Asian J Synth Nat Prod Chem*. 2023;1(1):11–15.

28. Chow EKH, Ho D. Cancer nanomedicine: from drug delivery to imaging. *Sci Transl Med*. 2013;5(216):216rv4. doi:10.1126/scitranslmed.3005872

29. Wang Y, Yang T, He Q. Strategies for engineering advanced nanomedicines for gas therapy of cancer. *Natl Sci Rev*. 2020;7(9):1485–1512. doi:10.1093/nsr/nwaa034

30. Zhao R, Wang B, Yang X, Xiao Y, Wang X, Shao C, et al. A drug-free tumor therapy strategy: cancer-cell-targeting calcification. *Angew Chem Int Ed*. 2016;55(17):5225–5229. doi:10.1002/anie.201601364

31. Lin H, Chen Y, Shi J. Nanoparticle-triggered in situ catalytic chemical reactions for tumour-specific therapy. *Chem Soc Rev*. 2018;47(6):1938–1958. doi:10.1039/c7cs00471k

32. Fan K, Xi J, Fan L, Wang P, Zhu C, Tang Y, et al. In vivo guiding nitrogen-doped carbon nanozyme for tumor catalytic therapy. *Nat Commun*. 2018;9(1):1440. doi:10.1038/s41467-018-03903-8

33. Yadav RN, Shaikh AL, Das A, Ray D, Banik BK. Asymmetric synthesis of 3-pyrrole substituted β-lactams through p-toluene sulphonic acid-catalyzed reaction of azetidine-2,3-diones with hydroxyprolines. *Curr Organocatalysis*. 2022;9(4):337–345.

34. Das A, Yadav RN, Banik BK. A novel Baker's yeast-mediated microwave-induced reduction of racemic 3-keto-2-azetidinones: facile entry to optically active hydroxy β-lactam derivatives. *Curr Organocatalysis*. 2022;9(2):195–198.

35. Das A, Banik BK. Versatile synthesis of organic compounds derived from ascorbic acid. *Curr Organocatalysis*. 2022;9(1):14–33.

36. Das A, Banik BK. Green Synthesis of Biologically Active pyrroles and related substrates via C-H Functionalization. In: Banik BK, ed. *Green Approaches in Medicinal Chemistry for Sustainable Drug Design*. Elsevier; 2024:101–131.

37. Das A, Banik BK. Sustainable reactions in the synthesis of heterocycles. *Curr Organocatalysis*. 2022;9(1):3–3. doi:10.2174/221333720901220328164523

38. Das A, Banik BK. 15 - Versatile Thiosugars in Medicinal Chemistry. In: Banik BK, ed. *Green Approaches in Medicinal Chemistry for Sustainable Drug Design*. Advances in Green Chemistry. Elsevier; 2020:549–574. doi:10.1016/B978-0-12-817592-7.00015-0

39. Das A, Banik BK. Chapter 18 - Versatile Thiosugars in Medicinal Chemistry. In: Banik BK, ed. *Green Approaches in Medicinal Chemistry for Sustainable Drug Design (Second Edition)*. Vol 2. Advances Green and Sustainable Chemistry. Elsevier; 2024:409–441. doi:10.1016/B978-0-443-16164-3.00018-2

40. Das A, Banik RNY. Ascorbic acid-mediated reactions in organic synthesis. *Curr Organocatalysis.* 2020;7(3):212–241.
41. Das A, Banik BK. Chapter 2 - Graphene Oxide and Modified Graphene Oxide-Mediated Synthesis of Medicinally Active Compounds. In: Banik BK, ed. *Green Approaches in Medicinal Chemistry for Sustainable Drug Design (Second Edition).* Vol 1. Advances in Green and Sustainable Chemistry. Elsevier; 2024:13–44. doi:10.1016/B978-0-443-16166-7.00002-5
42. Yadav RN, Hossain MdF, Das A, Srivastava AK, Banik BK. Organocatalysis: a recent development on stereoselective synthesis of o-glycosides. *Catal Rev.* 2024;66(1):1–118. doi:10.1080/01614940.2022.2041303
43. Zhao B, Wang Y, Yao X, Chen D, Fan M, Jin Z, et al. Photocatalysis-mediated drug-free sustainable cancer therapy using nanocatalyst. *Nat Commun.* 2021; 12(1):1345. doi:10.1038/s41467-021-21618-1
44. Zhang B, Wang X, Cheng Y. Photochromic immunoassay for tumor marker detection based on ZnO/AgI nanophotocatalyst. *Microchim Acta.* 2022; 189(2):77. doi:10.1007/s00604-021-05050-2
45. Piyarathne NS, Rasnayake RMSGK, Angammana R, Chandrasekera P, Ramachandra S, Weerasekera M, et al. Diagnostic salivary biomarkers in oral cancer and oral potentially malignant disorders and their relationships to risk factors – a systematic review. *Expert Rev Mol Diagn.* 2021;21(8):789–807. doi: 10.1080/14737159.2021.1944106
46. van der Waal I, Axéll T. Oral leukoplakia: a proposal for uniform reporting. *Oral Oncol.* 2002;38(6):521–526. doi:10.1016/s1368-8375(01)00125-7
47. Lin L, Song C, Wei Z, Zou H, Han S, Cao Z, et al. Multifunctional photo-dynamic/photothermal nano-agents for the treatment of oral leukoplakia. *J Nanobiotechnology.* 2022;20(1):106. doi:10.1186/s12951-022-01310-2
48. Duan X, Chan C, Guo N, Han W, Weichselbaum RR, Lin W. Photodynamic therapy mediated by nontoxic core–shell nanoparticles synergizes with immune checkpoint blockade to elicit antitumor immunity and antimetastatic effect on breast cancer. *J Am Chem Soc.* 2016;138(51):16686–16695. doi:10.1021/jacs. 6b09538
49. Wang P, Wu W, Gao R, Zhu H, Wang J, Du R, et al. Engineered cell-assisted photoactive nanoparticle delivery for image-guided synergistic photodynamic/ photothermal therapy of cancer. *ACS Appl Mater Interfaces.* 2019;11(15): 13935–13944. doi:10.1021/acsami.9b00022
50. Lan G, Ni K, Xu Z, Veroneau SS, Song Y, Lin W. Nanoscale metal–organic framework overcomes hypoxia for photodynamic therapy primed cancer immu-notherapy. *J Am Chem Soc.* 2018;140(17):5670–5673. doi:10.1021/jacs.8b01072
51. Ding C, Wang X, Song K, Zhang B, Wang J, Zhao Z, et al. Visible light enabled colorimetric tumor marker detection using ternary GO-C3N4-AgBr heterojunction nanophotocatalyst. *Sens Actuators B Chem.* 2018;268:376–382. doi:10.1016/j.snb.2018.04.146
52. Das A, Banik BK. Tellurium-based solar cells. *Phys Sci Rev.* 2023;8(12):4631–4658. doi:10.1515/psr-2021-0110
53. Das A, Banik BK. Semiconductor characteristics of tellurium and its imple-mentations. *Phys Sci Rev.* 2023;8(12):4659–4687. doi:10.1515/psr-2021-0108
54. Das A, Das A, Banik BK. Tellurium-based chemical sensors. *Phys Sci Rev.* 2023;8(12):4461–4501. doi:10.1515/psr-2021-0116

55. Das A, Das A, Banik BK. 9 Tellurium-Based Chemical Sensors. In: Krishna Banik B, Bajpai S, eds. *Tellurium Chemistry*. De Gruyter; 2022:183–224. doi:10.1515/9783110735840-009

56. Das A, Ray D, Banik BK. 4 Tellurium in Carbohydrate Synthesis. In: Krishna Banik B, Bajpai S, eds. *Tellurium Chemistry*. De Gruyter; 2022:85–106. doi:10.1515/9783110735840-004

57. Aldawood SAA, Das A, Banik BK. 11 Tellurium-Induced Cyclization of Olefinic Compounds. In: Krishna Banik B, Bajpai S, eds. *Tellurium Chemistry*. De Gruyter; 2022:249–290. doi:10.1515/9783110735840-011

58. Ray D, Das A, Mazumdar S, Banik BK. 12 Tellurium-Induced Functional Group Activation. In: Krishna Banik B, Bajpai S, eds. *Tellurium Chemistry*. De Gruyter; 2022:291–308. doi:10.1515/9783110735840-012

59. Das A, Banik BK. 5 Tellurium-Based Solar Cells. In: Krishna Banik B, Bajpai S, eds. *Tellurium Chemistry*. De Gruyter; 2022:107–134. doi:10.1515/9783110735840-005

60. Das A, Banik BK. 3 Semiconductor Characteristics of Tellurium and Its Implementations. In: Krishna Banik B, Bajpai S, eds. *Tellurium Chemistry*. De Gruyter; 2022:55–84. doi:10.1515/9783110735840-003

61. Das A, Ray D, Banik BK. Tellurium in carbohydrate synthesis. *Phys Sci Rev.* 2023;8(11):4157–4178. doi:10.1515/psr-2021-0109

62. Ray D, Das A, Mazumdar S, Banik BK. Tellurium-induced functional group activation. *Phys Sci Rev.* 2023;8(12):4821–4838. doi:10.1515/psr-2021-0221

63. Aldawood SAA, Das A, Banik BK. Tellurium-induced cyclization of olefinic compounds. *Phys Sci Rev.* 2023;8(12):4569–4609. doi:10.1515/psr-2021-0119

64. Zhen W, Liu Y, Jia X, Wu L, Wang C, Jiang X. Reductive surfactant-assisted one-step fabrication of a BiOI/BiOIO3 heterojunction biophotocatalyst for enhanced photodynamic theranostics overcoming tumor hypoxia. *Nanoscale Horiz.* 2019;4(3):720–726. doi:10.1039/C8NH00440D

65. Cheng Y, Kong X, Chang Y, Feng Y, Zheng R, Wu X, et al. Spatiotemporally synchronous oxygen self-supply and reactive oxygen species production on z-scheme heterostructures for hypoxic tumor therapy. *Adv Mater.* 2020;32(11):1908109. doi:10.1002/adma.201908109

66. Deng Z, Li H, Chen S, Wang N, Liu G, Liu D, et al. Near-infrared-activated anticancer platinum(IV) complexes directly photooxidize biomolecules in an oxygen-independent manner. *Nat Chem.* 2023;15(7):930–939. doi:10.1038/s41557-023-01242-w

67. Sigel A, Sigel H, Freisinger E, Sigel RKO. *Metallo-Drugs: Development and Action of Anticancer Agents*. Walter de Gruyter GmbH & Co KG; 2018.

68. Meier-Menches SM, Gerner C, Berger W, Hartinger CG, Keppler BK. Structure-activity relationships for ruthenium and osmium anticancer agents – towards clinical development. *Chem Soc Rev.* 2018;47(3):909–928. doi:10.1039/c7cs00332c

69. Knoll JD, Turro C. Control and utilization of ruthenium and rhodium metal complex excited states for photoactivated cancer therapy. *Coord Chem Rev.* 2015;282-283:110–126. doi:10.1016/j.ccr.2014.05.018

70. Heinemann F, Karges J, Gasser G. Critical overview of the use of Ru(II) polypyridyl complexes as photosensitizers in one-photon and two-photon photodynamic therapy. *Acc Chem Res.* 2017;50(11):2727–2736. doi:10.1021/acs.accounts.7b00180

71. Spring BQ, Rizvi I, Xu N, Hasan T. The role of photodynamic therapy in overcoming cancer drug resistance. *Photochem Photobiol Sci Off J Eur Photochem Assoc Eur Soc Photobiol.* 2015;14(8):1476–1491. doi:10.1039/c4pp00495g

72. Teicher BA. Hypoxia and drug resistance. *Cancer Metastasis Rev.* 1994;13(2):139–168. doi:10.1007/BF00689633

73. Höckel M, Vaupel P. Tumor hypoxia: definitions and current clinical, biologic, and molecular aspects. *J Natl Cancer Inst.* 2001;93(4):266–276. doi:10.1093/jnci/93.4.266

74. Huang H, Banerjee S, Qiu K, Zhang P, Blacque O, Malcomson T, et al. Targeted photoredox catalysis in cancer cells. *Nat Chem.* 2019;11(11):1041–1048. doi:10.1038/s41557-019-0328-4

75. Adorno-Cruz V, Kibria G, Liu X, Doherty M, Junk DJ, Guan D, et al. Cancer stem cells: targeting the roots of cancer, seeds of metastasis, and sources of therapy resistance. *Cancer Res.* 2015;75(6):924–929. doi:10.1158/0008-5472.CAN-14-3225

76. Chen J, Cao X, An Q, Zhang Y, Li K, Yao W, et al. Inhibition of cancer stem cell like cells by a synthetic retinoid. *Nat Commun.* 2018;9(1):1406. doi:10.1038/s41467-018-03877-7

77. Rivnay J, Wang H, Fenno L, Deisseroth K, Malliaras GG. Next-generation probes, particles, and proteins for neural interfacing. *Sci Adv.* 2017;3(6):e1601649. doi:10.1126/sciadv.1601649

78. Deubner J, Coulon P, Diester I. Optogenetic approaches to study the mammalian brain. *Curr Opin Struct Biol.* 2019;57:157–163. doi:10.1016/j.sbi.2019.04.003

79. Yu Y, Yang X, Reghu S, Kaul SC, Wadhwa R, Miyako E. Photothermogenetic inhibition of cancer stemness by near-infrared-light-activatable nanocomplexes. *Nat Commun.* 2020;11(1):4117. doi:10.1038/s41467-020-17768-3

80. Wang Q, Ji Y, Shi J, Wang L. NIR-Driven water splitting H2 production nanoplatform for H_2-mediated Cascade-amplifying synergetic cancer therapy. *ACS Appl Mater Interfaces.* 2020;12(21):23677–23688. doi:10.1021/acsami.0c03852

81. Sang D, Wang K, Sun X, Wang Y, Lin H, Jia R, et al. NIR-Driven intracellular photocatalytic O_2 evolution on Z-scheme Ni3S2/Cu1.8S@HA for hypoxic tumor therapy. *ACS Appl Mater Interfaces.* 2021;13(8):9604–9619. doi:10.1021/acsami.0c21284

82. Liu P, Yang W, Shi L, Zhang H, Xu Y, Wang P, et al. Concurrent photothermal therapy and photodynamic therapy for cutaneous squamous cell carcinoma by gold nanoclusters under a single NIR laser irradiation. *J Mater Chem B.* 2019;7(44):6924–6933. doi:10.1039/C9TB01573F

83. Chen Q, Chen J, Yang Z, Zhang L, Dong Z, Liu Z. NIR-II light activated photodynamic therapy with protein-capped gold nanoclusters. *Nano Res.* 2018;11(10):5657–5669. doi:10.1007/s12274-017-1917-4

84. Wang J, Liu D, Zhu Y, Zhou S, Guan S. Supramolecular packing dominant photocatalytic oxidation and anticancer performance of PDI. *Appl Catal B Environ.* 2018;231:251–261. doi:10.1016/j.apcatb.2018.03.026

85. Zhang Z, Wang L, Liu W, Yan Z, Zhu Y, Zhou S, et al. Photogenerated-hole-induced rapid elimination of solid tumors by the supramolecular porphyrin photocatalyst. *Natl Sci Rev.* 2021;8(5):nwaa155. doi:10.1093/nsr/nwaa155

86. Yu YA, Shabahang S, Timiryasova TM, Zhang Q, Beltz R, Gentschev I, et al. Visualization of tumors and metastases in live animals with bacteria and vaccinia virus encoding light-emitting proteins. *Nat Biotechnol.* 2004;22(3):313–320. doi:10.1038/nbt937

87. Luo CH, Huang CT, Su CH, Yeh CS. Bacteria-mediated hypoxia-specific delivery of nanoparticles for tumors imaging and therapy. *Nano Lett.* 2016; 16(6):3493–3499. doi:10.1021/acs.nanolett.6b00262

88. Forbes NS. Engineering the perfect (bacterial) cancer therapy. *Nat Rev Cancer.* 2010;10(11):785–794. doi:10.1038/nrc2934

89. Summers DK. The kinetics of plasmid loss. *Trends Biotechnol.* 1991;9(8):273–278. doi:10.1016/0167-7799(91)90089-z

90. Zitvogel L, Daillère R, Roberti MP, Routy B, Kroemer G. Anticancer effects of the microbiome and its products. *Nat Rev Microbiol.* 2017;15(8):465–478. doi:10.1038/nrmicro.2017.44

91. Sakimoto KK, Wong AB, Yang P. Self-photosensitization of nonphotosynthetic bacteria for solar-to-chemical production. *Science.* 2016;351(6268): 74–77. doi:10.1126/science.aad3317

92. Liu C, Gallagher JJ, Sakimoto KK, Nichols EM, Chang CJ, Chang MCY, et al. Nanowire–bacteria hybrids for unassisted solar carbon dioxide fixation to value-added chemicals. *Nano Lett.* 2015;15(5):3634–3639. doi:10.1021/acs.nanolett.5b01254

93. Zheng DW, Chen Y, Li ZH, Xu L, Li CX, Li B, et al. Optically-controlled bacterial metabolite for cancer therapy. *Nat Commun.* 2018;9(1):1680. doi:10.1038/s41467-018-03233-9

94. Bambou JC, Giraud A, Menard S, Begue B, Rakotobe S, Heyman M, et al. In vitro and ex vivo activation of the TLR5 signaling pathway in intestinal epithelial cells by a commensal *Escherichia coli* strain. *J Biol Chem.* 2004;279(41):42984–42992. doi:10.1074/jbc.M405410200

95. Zheng DW, Li B, Li CX, Fan JX, Lei Q, Li C, et al. carbon-dot-decorated carbon nitride nanoparticles for enhanced photodynamic therapy against hypoxic tumor via water splitting. *ACS Nano.* 2016;10(9):8715–8722. doi:10.1021/acsnano.6b04156

96. Marletta MA, Yoon PS, Iyengar R, Leaf CD, Wishnok JS. Macrophage oxidation of L-arginine to nitrite and nitrate: nitric oxide is an intermediate. *Biochemistry.* 1988;27(24):8706–8711. doi:10.1021/bi00424a003

97. Benson S, de Moliner F, Fernandez A, Kuru E, Asiimwe NL, Lee JS, et al. Photoactivatable metabolic warheads enable precise and safe ablation of target cells in vivo. *Nat Commun.* 2021;12(1):2369. doi:10.1038/s41467-021-22578-2

Index